Seafoods and Fish Oils
in Human Health and Disease

FOOD SCIENCE AND TECHNOLOGY

A Series of Monographs, Textbooks, and Reference Books

Editorial Board

Owen R. Fennema
University of Wisconsin—Madison

Gary W. Sanderson
Universal Foods Corporation

Pieter Walstra
Wageningen Agricultural University

Marcus Karel
Massachusetts Institute of Technology

Steven R. Tannenbaum
Massachusetts Institute of Technology

John R. Whitaker
University of California—Davis

1. Flavor Research: Principles and Techniques, *R. Teranishi, I. Hornstein, P. Issenberg, and E. L. Wick (out of print)*
2. Principles of Enzymology for the Food Sciences, *John R. Whitaker*
3. Low-Temperature Preservation of Foods and Living Matter, *Owen R. Fennema, William D. Powrie, and Elmer H. Marth*
4. Principles of Food Science
   Part I: Food Chemistry, *edited by Owen R. Fennema*
   Part II: Physical Methods of Food Preservation, *Marcus Karel, Owen R. Fennema, and Daryl B. Lund*
5. Food Emulsions, *edited by Stig Friberg*
6. Nutritional and Safety Aspects of Food Processing, *edited by Steven R. Tannenbaum*
7. Flavor Research: Recent Advances, *edited by R. Teranishi, Robert A. Flath, and Hiroshi Sugisawa*
8. Computer-Aided Techniques in Food Technology, *edited by Israel Saguy*
9. Handbook of Tropical Foods, *edited by Harvey T. Chan*
10. Antimicrobials in Foods, *edited by Alfred Larry Branen and P. Michael Davidson*
11. Food Constituents and Food Residues: Their Chromatographic Determination, *edited by James F. Lawrence*

12. Aspartame: Physiology and Biochemistry, *edited by Lewis D. Stegink and L. J. Filer, Jr.*
13. Handbook of Vitamins: Nutritional, Biochemical, and Clinical Aspects, *edited by Lawrence J. Machlin*
14. Starch Conversion Technology, *edited by G. M. A. van Beynum and J. A. Roels*
15. Food Chemistry: Second Edition, Revised and Expanded, *edited by Owen R. Fennema*
16. Sensory Evaluation of Food: Statistical Methods and Procedures, *Michael O'Mahony*
17. Alternative Sweeteners, *edited by Lyn O'Brien Nabors and Robert C. Gelardi*
18. Citrus Fruits and Their Products: Analysis and Technology, *S. V. Ting and Russell L. Rouseff*
19. Engineering Properties of Foods, *edited by M. A. Rao and S. S. H. Rizvi*
20. Umami: A Basic Taste, *edited by Yojiro Kawamura and Morley R. Kare*
21. Food Biotechnology, *edited by Dietrich Knorr*
22. Food Texture: Instrumental and Sensory Measurement, *edited by Howard R. Moskowitz*
23. Seafoods and Fish Oils in Human Health and Disease, *John E. Kinsella*

Other Volumes in Preparation

Seafoods and Fish Oils in Human Health and Disease

John E. Kinsella

Institute of Food Science
Cornell University
Ithaca, New York

MARCEL DEKKER, INC.   New York and Basel

Library of Congress Cataloging-in-Publication Data

Kinsella, John E. [date]
　　Seafoods and fish oils in human health and disease.

　　(Food science and technology (Marcel Dekker, Inc., ; 23))
　　Includes bibliographies and index.
　　1. Fish oils in human nutrition. 2. Unsaturated fatty acids in human nutrition. 3. Seafoods--Composition.
I. Title. II. Series. [DNLM: 1. Fatty Acids, Unsaturated.
2. Fish Oils. 3. Fishes. W1 F0509P v.23 QU 90 K56s]
QP752.F57K56　1987　613.2'8　87-5264
ISBN 0-6247-7771-9

Copyright © 1987 by MARCEL DEKKER, INC.
All Rights Reserved

Neither this book nor any part may be reproduced or transmitted in any form or by any means, electronic or mechanical, including photocopying, microfilming, and recording, or by any information storage and retrieval system, without permission in writing from the publisher.

MARCEL DEKKER, INC.
270 Madison Avenue, New York, New York 10016

Current printing (last digit):
10　9　8　7　6　5　4　3　2　1

PRINTED IN THE UNITED STATES OF AMERICA

Preface

Dietary components have been implicated in many major public health diseases, particularly as risk factors in heart disease, atherosclerosis, cancer, and allergies. Modifications of dietary habits, particularly the reduction of fat and cholesterol consumption and incresed ingestion of polyunsaturated fatty acids (PUFAs) of the n-6 (ω-6) family, has been strongly advocated. While these modifications may have contributed to the reduction of serum lipids, cholesterol, and heart disease, there is some concern that consumption of n-6 PUFAs may be excessive and as a result tissues may be overenriched in arachidonic acid. This may accentuate the production of eicosanoids in response to normal stimuli and facilitate or exacerbate pathophysiological conditions such as tumor growth, arthritis, and conceivably thrombosis [1].

While much is known about nutrient needs and the minimum requirements to avoid frank deficiencies, more information about nutrient interrelationships and optimum balance is needed, especially for the dietary unsaturated fatty acids. For example, the apparent requirement for the essential fatty acid linoleic acid (18:2n-6) is about 1% of caloric intake; however, current levels of consumption may be within the 5 to 8 en% range. Although linoleic acid consumption in this range may depress serum cholesterol, the enhanced levels of arachidonic acid in tissue may oversensitize responses to or override normal controls of eicosanoids and thus precipitate some chronic and acute disease states [1,2,4].

Preface

Recent evidence suggests that unsaturated fatty acids of the n-3 (ω-3) configuration, by competing with n-6 fatty acids, may modify the effects of n-6 fatty acids, thereby ameliorating the undesirable effects of excessive eicosanoid production. This evidence reinforces the need for more information concerning the optimum intake and ratio of unsaturated fatty acids of the n-3 and n-6 structures and the potential beneficial effects of fish or fish oil consumption on public health. Fish oil consumption may provide a possible dietary approach for reducing a number of pathophysiological states [1-4].

In this volume some specific information concerning the effects of fish and fish oils on lipid metabolism in relation to some public health problems are reviewed, based on the recent literature. Since fish oils may be of value for dietary intervention, the sources, availability, processing, and safety of fish oils are briefly discussed. Finally the characteristics of some fish and the fatty acid contents of common edible species are presented.

This volume is not intended to be a critical nor comprehensive review of marine oils or seafoods. The purpose is to provide an update of contemporary knowledge, indicate areas where information is needed, and thereby facilitate decisions about appropriate intervention strategies. Fish oil is not a cure-all, but it may be an effective and desirable dietary component when included in a prudent, low-fat diet with a balanced polyunsaturated fatty acid composition.

This book represents the dedicated efforts of several people, especially Pauline Herold, who set the pace initially, Nina Cardillo, who painstakingly helped with tables and library work, and Judy Stewart, who edited and assembled the final draft.

The support of the New York Sea Grant Program and the encouragement of Donald Squires are gratefully acknowledged. The stimulation of many discussions with colleagues and co-workers, especially Bruce German, Belur Lokesh, Geza Bruckner, and Joy Swanson greatly helped in this effort, and finally, the support and patience of my wife Ruth Ann and family are gratefully appreciated.

John E. Kinsella

Preface

REFERENCES

1. Lands, W. E. M. (1985). *Fish and Human Health*. Academic Press, New York.
2. Lands, W. E. M. (1986). Renewed questions about polyunsaturated fatty acids. *Nutrition Rev*. 44:189-95.
3. Dyerberg, J. (1986). Linolenate derived polyunsaturated fatty accids and prevention of atherosclerosis. *Nutrition Rev*. 4:126-36.
4. Simopoulas, A.P., Kifer, R.R., and Martin, R.E. (1986). *Health Effects of Polyunsaturated Fatty Acids in Seafoods*. Academic Press, New York.

Contents

Preface   iii

Abbreviations   xi

1. DIETARY FATS AND CARDIOVASCULAR DISEASE   1

    Atherosclerosis   1
    Risk Factors   3
    Plasma Lipid Changes Associated with Diets Rich in
      Cholesterol and Saturated Fat   4
    Plasma Lipid Changes Associated with Diets Rich in
      Polyunsaturated Fats   5
    Dietary Polyunsaturated Fatty Acids   6
    Dietary Fat and Platelet Function   7
    Essential Fatty Acids   9
    Recent Trends in Mortality Due to Cardiovascular
      Diease and Their Possible Causes   12
    Studies on Greenland Eskimos   13
    Studies on Other Populations   19
    References   20

2. EICOSANOID METABOLISM: GENERAL BACKGROUND
    AND OVERVIEW   25

    Introduction   25
    Arachidonic Acid   27
    Eicosanoid Synthesis   28
    Conclusion   36
    References   37

3. THE EFFECTS OF OMEGA-3 POLYUNSATURATED FATTY
    ACID CONSUMPTION ON THE PLASMA, PLATELET,
    VESSEL WALL, AND ERYTHROCYTE CHARACTERISTICS
    OF HUMAN SUBJECTS IN FEEDING TRIALS   41

Introduction ... 41
Effects of Fatty Fish, Marine Oil, or n-3 PUFA
    Consumption on Plasmatic Characteristics ... 68
Comparison of the Findings from Feeding Trials
    and the Plasmatic Characteristics of Eskimos ... 79
Effects of Fatty Fish, Marine Oil, or n-3 PUFA
    Consumption on Platelet Characteristics ... 79
Comparison of the Findings from Feeding Trials and
    the Platelet Characteristics of Eskimos ... 90
Effects of Fatty Fish, Marine Oil, or n-3 PUFA
    Consumption on the Production of Prostaglandins
    I_2 and I_3 by Vessel Walls and on Levels of
    Urinary Prostaglandin Metabolites ... 90
Effects of Fatty Fish, Marine Oil, or n-3 PUFA
    Consumption on Erythrocyte Characteristics ... 92
Effects of Fatty Fish, Marine Oil, or n-3 PUFA
    Consumption on Other Blood Characteristics ... 94
Discussion ... 96
References ... 100

4. **THE EFFECTS OF DIETARY n-3 PUFAs OF FISH OILS ON SERUM LIPIDS, EICOSANOIDS, AND THROMBOTIC EVENTS: OBSERVATIONS FROM ANIMAL FEEDING TRIALS** ... 107

Introduction ... 107
Feeding Trials ... 107
Effect of Marine Oil or n-3 PUFA Consumption on
    Plasma Characteristics ... 113
Effects of Marine Oil or n-3 PUFA Consumption on
    Platelet Characteristics ... 126
Effects of Marine Oil or n-3 PUFA Consumption on
    Cardiac Characteristics ... 133
Effects of Marine Oil and n-3 PUFA Consumption on
    Aortic Tissue ... 138
Effects of Marine Oil or n-3 PUFA Consumption on
    Other Blood Characteristics ... 141
Effects of Marine Oil or n-3 PUFA Consumption on
    Hepatic Characteristics ... 144
Effects of Marine Oil or n-3 PUFA Consumption on
    Lung Lipids ... 150
Effects of Marine Oil or n-3 PUFA Consumption on
    Kidney Lipids ... 153
Effects of Marine Oil and m-3 PUFA Consumption on
    Adipose Tissue ... 155
Effects of Marine Oil and n-3 PUFA Consumption on
    Cerebral and Neurological Characteristics ... 156
Discussion ... 159
References ... 161

Contents

5.	**DIETARY POLYUNSATURATED FATTY ACIDS AND CANCER**	165
	Introduction	165
	Experimental Evidence	166
	Discussion	168
	References	168
6.	**FISH AND n-3 POLYUNSATURATED FATTY ACID CONSUMPTION IN THE UNITED STATES: SOME CALCULATIONS**	171
	Introduction	171
	The n-3 PUFA Content of Fish	171
	n-3 PUFA Intake with the Current U.S. Seafood Consumption	172
	n-3 PUFA Intake with a Hypothetical Higher Consumption	173
	Comparison to the n-3 PUFA Intake of Eskimos and Feeding Trial Participants	174
	References	175
7.	**CHOLESTEROL AND FAT SOLUBLE VITAMINS IN FISH LIPIDS**	177
	Cholesterol Content of Fish Oils	177
	Vitamin A and D Content of Fish Oils	178
	Vitamin E Content of Fish Oils	186
	References	189
8.	**COMPONENTS AFFECTING THE SAFETY OF FISH OILS**	193
	Introduction	193
	The Autoxidation of Fish Oils	193
	The Effects of Monoenoic Acids: Erucic Acid	197
	The Presence of Organochlorine Pesticides and Polychlorinated Biphenyls	199
	Heavy Metals: Lead, Mercury, and Cadmium	203
	References	204
9.	**EDIBLE FISH OIL PROCESSING AND TECHNOLOGY**	209
	Introduction	209
	Catch	209
	Production	210
	Processing	213
	Problems Involved in Processing	219
	Advances in Processing of Fish Oils	222
	Quality Control Standards for Fish Oils	223
	Storage and Packaging of Fish Oils	224

	Marketing of Fish Oils	225
	Industrial Needs	225
	References	226
10.	SUMMARY OF THE HEALTH IMPLICATIONS OF DIETARY FISH AND FISH OILS AND RESEARCH NEEDS	231
	Health Effects of Fish Oils	231
	Research Needs	235
	References	237
11.	POTENTIAL SOURCES OF FISH OIL: FATTY FISH IN U.S. WATERS	239
	Fatty Fish in U.S. Waters	239
	Common Fish Species	262
	References	295

Index   *301*

Abbreviations

AA	Arachidonic acid (20:4n-6)
ADP	Adenosine diphosphate
AT III	Antithrombin III
CLO	Cod liver oil
DG	Diacylglyceride
DHA	Docosahexaenoic acid (22:6n-3)
EPA	Eicosapentaenoic acid (20:5n-3)
FA	Fatty acid
HDL	High density lipoprotein
HEPA	(5Z,8Z,10E,14Z,17Z)-12-L-hydroxyeicosapentaenoic acid
HETE	(5Z,8Z,10E,14Z)-12-L-hydroxyeicosatetraenoic acid
HHT	(5Z,8E,10E)-12-L-hydroxyheptadecatrienoic acid
HHTE	(5Z,8E,10E,14Z)-12-L-hydroxyheptadecatetraenoic acid
LA	Linoleic acid (18:2n-6)
LDL	Low density lipoprotein
LT	Leukotriene
MDA	Malondialdehyde
MG	Monoglyceride
PC	Phosphatidylcholine
PCB	Polychlorinated biphenyl
PE	Phosphatidylethanolamine
PI	Phosphatidylinositol
PDGF	Platelet-derived growth factor
PG	Prostaglandin

PGE2	Prostaglandin E2
PGG2	Prostaglandin endoperoxide
PGH2	Prostaglandin endoperoxide
PGI_2	Prostacyclin
PGI_3	Prostaglandin I_3
PRP	Platelet rich plasma
PUFA	Polyunsaturated fatty acid
TBA	Thiobarbituric acid
TXA_2	Thromboxane
TXA_3	Thromboxane A_3
TXB_3	Thromboxane B_3
VLDL	Very low density lipoprotein
ω	Omega: used interchangeably with n to designate position of double bond in polyunsaturated fatty acids.

Seafoods and Fish Oils
in Human Health and Disease

1
Dietary Fats and Cardiovascular Disease

Cardiovascular diseases are the leading causes of death in western industrialized nations and account for nearly half of all deaths in the United States [1]. Among cardiovascular diseases, those resulting from thromboembolic complications such as coronary occlusion, infarction, and stroke are the main cause of death and disability. The direct and indirect costs attributed to cardiovascular diseases have been estimated to exceed $80 billion annually [2].

ATHEROSCLEROSIS

Atherosclerosis is a primary cause of cardiovascular disease [2]. Localized thickenings or plaques of fatty and fibrous material on arterial walls cause narrowing of the vessel lumen. Clumps of aggregated platelets called thrombi may form at the sites of atherosclerotic lesions, and impaired circulation and ischemia occur when atherosclerotic lesions and thrombi obstruct the flow of blood [3].

Three distinct layers make up arterial walls. The intima, or innermost layer bordering the lumen, consists of flattened endothelial cells. A fenestrated membrane composed largely of elastin separates the intima from the media or middle layer. This is formed of concentric wrappings of smooth muscle cells, elastin, and fibrous material supported in a matrix of glycoproteins, proteoglycans, and glucosaminoglycans. The adventitia, or outermost layer, contains fibroblasts and smooth muscle cells loosely arranged among bundles of collagen fibers. The outer half of the arterial wall is supplied by the capillary network of the vasa vasorum while the inner half relies on diffusion from the lumen for its nutriment [4,5,6].

Injury to the endothelial cells of the arterial wall initiates atherosclerosis [5]. Hypertension has been suggested as a cause. Its action may be purely mechanical, causing damage by stretching the endothelial cells. Vasoactive compounds in the blood of hypertensive individuals have also been implicated [7].

Injury to the endothelial cells increases their permeability. This allows the entry of cholesterol from circulating lipoproteins into the deeper layers of the arterial wall [5]. Lipoproteins, which are macromolecular complexes synthesized by the liver for the transport of cholesterol and triglycerides, are also composed of phospholipids and specific proteins termed apoproteins. The various classes of lipoproteins, which include very low density (VLDL), low density (LDL), and high density (HDL) lipoproteins, differ in size and relative proportion of constituent components (triglycerides, cholesterol, phospholipids) [37].

Another response to injury is a thickening of the arterial wall due to the proliferation of smooth muscle cells. Areas of necrosis may develop when the thickening of the arterial walls limits the oxygen available to areas not supplied by the vasa vasorum [5]. The source of cholesterol in atherosclerotic lesions is believed to be the low density lipoproteins (LDLs) [8,9,10,11]. Lipid droplets are deposited and can form a dense coating on elastin and collagen fibers. Fat-filled cells derived from smooth muscle cells or circulating monocytes also appear [8,11].

More recent evidence points to an important role for platelet-vessel wall interactions following endothelial damage in the development of atherosclerosis. If endothelial injury has led to desquamation, platelets may adhere to the exposed surfaces and release platelet-derived growth factor (PDGF), a low molecular weight glycoprotein that induces the migration of smooth muscle cells to the intima and their proliferation while also stimulating LDL-receptor activity [11,12,13]. The proliferation of smooth muscle cells is also thought to be a response to the entry of circulating monocytes and macrophages, which ingest lipoproteins, triglycerides, and cholesterol and release a potent mitogen. They also release chemoattractants that recruit additional cells which adhere [12,13]. In time these become invested by smooth muscle cells, and with further progression of the lesions, fibrous plaques with a surface layer of collagen are formed. Collagen, which stimulates platelet and macrophage adhesion, can constitute 40 to 60% of the dry mass of the lesions [5]. Calcification may also occur subsequently with the deposition of hydroxyapatite crystals [5]. The formation of thrombi at the sites of atherosclerotic lesions may follow. Fissures in the plaques lead to hemorrhage and the activation and aggregation of platelets [14]. Stasis in the narrowed artery or the hypercoagulability of blood in the vessel may also cause the formation of thrombi [5]. Thrombi and occlusion are major causes of cardiac and cerebral infarction [38].

The important role of platelets and macrophages in the etiology of atherogenesis has been documented by Ross [39]. Eicosanoids may be intimately involved in this process via their activity in enhancing the adhesiveness of both platelets and macrophages. The possibilities that excessive eicosanoid production may promote plaque development leading to atherosclerosis and that diets high in n-3 PUFAs may minimize this effect warrant careful study.

RISK FACTORS

Several risk factors have been implicated in the etiology of atherosclerosis and ischemic heart disease. A prospective epidemiological study carried out since 1949 among more than 5,000 residents of Framingham, Massachusetts, identified numerous factors that increase the risk of cardiovascular disease [15]. These include the levels of various plasma constituents, especially lipoproteins, and the lifestyle patterns, personality traits, and genetic background of individuals [15,16].

It was observed that total plasma cholesterol and VLDL and LDL concentrations have positive associations with subsequent incidence of heart disease, whereas HDL concentrations are negatively associated with risk. Risk is heightened by an increase in the ratio of total to HDL cholesterol [15].

Increased plasma cholesterol and lipoprotein levels may accelerate the development of arterial disease by increasing the quantity of circulating lipid available for incorporation into arterial walls [8]. Two possible mechanisms have been suggested for the apparent beneficial effect of HDLs. HDLs may reduce the arterial uptake of cholesterol carried by other lipoproteins or interfere with the binding of lipoproteins to arterial walls. They may also remove cholesterol from arterial walls and transport it to the liver [8,10].

A statistical association between cardiovascular disease and increased blood pressure also has been observed [8,15]. Once the permeability of the arterial walls has been altered, hypertension may allow a larger volume of plasma and plasma lipoproteins to enter them, leading to greater lipid accumulation [8].

PLASMA LIPID CHANGES ASSOCIATED WITH DIETS RICH IN CHOLESTEROL AND SATURATED FAT

It is known from human feeding trials that an increase in dietary cholesterol is usually accompanied by an increase in the amount absorbed by the body. Increased absorption leads to higher levels of cholesterol in plasma and body pools [8]. Generally, greater amounts of cholesterol in the diet raise the concentration of LDLs, although the mechanism remains unclear. The influx of excess dietary cholesterol to the liver may increase the production of LDLs. Although most data suggest that these lipoproteins arise largely from the degradation of VLDLs, it has been proposed that the liver may also secrete LDLs or lipoproteins that are quickly converted to LDLs directly in the plasma. It also has been suggested

Dietary Fats and Cardiovascular Disease

that increased dietary cholesterol augments the formation of abnormally large LDLs with an extra content of cholesterol. Both mechanisms may facilitate the increased deposition of cholesterol in arterial walls [9].

PLASMA LIPID CHANGES ASSOCIATED WITH DIETS RICH IN POLYUNSATURATED FATS

In feeding trials, the replacement of saturated fats and cholesterol in the diet by polyunsaturated fats from vegetable sources caused changes that are associated with a decreased risk of cardiovascular disease [17]. Diets rich in saturated fat caused increases in plasma cholesterol; the consumption of polyunsaturated fats had the opposite effect [17, 9]. Equations have been developed that predict that on a gram for gram basis saturated fatty acids will raise cholesterol levels about two times as much as vegetable polyunsaturated acids will lower them [17]. The cholesterol lowering effect of polyunsaturated fats is related to the magnitude of the difference between P/S ratios (polyunsaturated to saturated fatty acids) of the saturated and polyunsaturated fat diets and to the presence of cholesterol. With larger differences in P/S ratios, a greater drop in cholesterol concentration can be expected. When cholesterol is absent from diets, plasma cholesterol levels are less responsive to changes in the dietary P/S ratio [17].

The consumption of polyunsaturated fats is also accompanied by marked reductions in LDL and VLDL concentrations [9,18]. The effect on HDL concentrations is variable and remains unclear [17,18].

The mechanisms by which polyunsaturated fats reduce plasma cholesterol concentrations are still poorly understood [9,17,18,19]. An increased fecal excretion of cholesterol and its derivatives often accompanies a fall in plasma cholesterol [9,17], perhaps representing the enhanced elimination of the cholesterol lost from the plasma compartment [9]. A number of other potential mechanisms has also been suggested: a decrease in cholesterol absorption in the gut; a reduction in its synthesis by the body; a shift in its content from the plasma to other body components; changes in the rates of synthesis or catabolism of individual lipoproteins; depressed hepatic synthesis of fatty acids and triglycerides for LDLs; or an increase in the polyunsaturated fatty acid content of LDLs, which alters their structure so they only can accommodate a smaller quantity of cholesterol [9,17,18].

The cause of the decreases in LDL and VLDL concentrations also remains the object of research. It was proposed that an increase in polyunsaturated fatty acid consumption changes the fatty acid composition and fluidity of lipoproteins, making them better substrates for catabolic enzymes or for binding to cellular receptors. Cellular membranes may be similarly affected, changing the activity of surface receptors and membrane-bound enzymes [17].

DIETARY POLYUNSATURATED FATTY ACIDS

Recognition of a correlation between dietary fat and cardiovascular disease has led to dietary recommendations that advocate replacing dietary saturated fats and cholesterol with polyunsaturated fatty acids (PUFAs). The principal families of unsaturated fatty acids are shown in Figure 1.1. Of major interest are the n-6 and n-3 families, characterized by the position of the first double bond (at the sixth and third carbon atom, respectively, counting from the methyl end of the carbon chain). N-6 unsaturated fatty acids are essential in the diet and occur most commonly in plant oils. Linoleic acid (18:2n-6), the major dietary source of the n-6 PUFAs, is abundant in vegetable oils (soybean, corn, and safflower) and is the precursor of arachidonic acid (20:4n-6).

The n-3 family occurs in small amounts in most foods except fish. Linolenic acid (18:3n-3) is present in low concentrations in leafy plant tissues and soybean oil. Eicosapentaenoic acid (EPA) (20:5n-3) and docosahexaenoic acid (DHA) (22:6n-3) originate in unicellular phytoplankton and seaweeds, and once incorporated into the lipids of fish and other marine animals that consume the algae, they are passed on through the food chain to other species [23]. Persons consuming a normal diet have low tissue levels of EPA and

Oleic 18:1 n-9
$$CH_3(CH_2)_7-CH=CH-(CH_2)_7COOH$$

Linoleic 18:2 n-6
$$CH_3-(CH_2)_4-CH=CH-CH_2-CH=CH-(CH_2)_7-COOH$$

Arachidonic 20:4 n-6
$$CH_3-(CH_2)_4-CH=CH-CH_2-CH=CH-CH_2-CH=CH-CH_2-CH=CH(CH_2)_3-COOH$$
$$\underset{n-6}{\underline{\qquad\qquad}}$$

Linolenic 18:3 n-3
$$CH_3-CH_2-CH=CH-CH_2-CH=CH-CH_2-CH=CH-(CH_2)_7-COOH$$

Eicosapentaenoic 20:5 n-3
$$CH_3-CH_2-CH=CH-CH_2-CH=CH_2-CH_2-CH=CH-CH_2-CH=CH-CH=CH-(CH_2)_3-COOH$$

Docosahexaenoic 22:6 n-3
$$CH_3-CH_2-CH=CH-CH_2-CH=CH-CH_2-CH=CH-CH_2-CH=CH-CH_2-CH=CH-CH_2-CH=CH-(CH_2)_2-COOH$$
$$\underset{n-3}{\underline{\qquad}}$$

FIGURE 1.1 Different families of unsaturated fatty acids.

DHA, but persons who regularly consume fish have increased tissue levels.

EPA and DHA also are derived from linolenic acid (Figure 1.2). Although linoleic acid competes with linolenic acid for the same enzymes in the desaturation and elongation reactions, the formation of arachidonic acid from linoleic acid is favored.

DIETARY FAT AND PLATELET FUNCTION

The effect of dietary fat on the development of cardiovascular disease may also be mediated through other, perhaps more important, mechanisms, including platelet function, hemodynamics, and the formation of thrombi. Arachidonic acid (20:4n-6), a twenty-carbon polyunsaturated fatty acid with four double bonds, occurs in the diet (i.e. meats) but is mainly synthesized in the liver from dietary

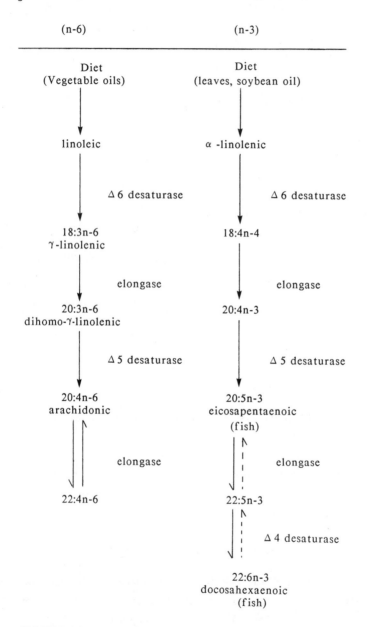

FIGURE 1.2 Metabolic pathways of the n-6 and n-3 PUFAs.

linoleic acid of vegetable origin (Figure 1.2). It is the precursor for the synthesis of eicosanoids (i.e. prostaglandins and leukotrienes), compounds which have profound effects on platelet aggregation, vascular muscle tone, and the diameter of vessel lumens. The prostanoid thromboxane, formed from arachidonic acid in platelets, is a potent proaggregatory agent that stimulates platelet adhesion and aggregation and induces vasoconstriction, thereby increasing blood pressure. This is counteracted by prostacyclin, a prostaglandin synthesized from arachidonic acid in vessel walls, which effectively inhibits platelet aggregation, induces vasodilation, and reduces blood pressure [17,20]. Since thromboxane and prostacyclin are ultimately derived from dietary polyunsaturated fatty acids, it is possible that manipulation of the fat content and composition of the diet may alter hemodynamics, platelet behavior, vessel wall characteristics, and the development of thrombi [17]. Dietary manipulation, by altering the amounts and ratios of bioactive thromboxane and prostacyclin in the vascular system, may also affect ischemic heart disease.

Based on the current understanding of thrombosis, reduction of proaggreegatory thromboxane synthesis, while maintaining or enhancing the anti-aggregatory prostacyclin production, is most desirable because prostacyclin inhibits platelet activation by both thromboxane-dependent and thromboxane-independent mechanisms. Within this context, EPA is of particular interest, not only because of its ability to inhibit thromboxane synthesis by platelets but also because it can be converted to a bioactive anti-aggregatory prostacyclin analog without significantly interfering with prostacyclin synthesis [38].

ESSENTIAL FATTY ACIDS

Specific dietary unsaturated fatty acids are required to prevent the multiple symptoms of essential fatty acid deficiency. Humans need specific essential fatty acids, such as linoleic acid and its elongated, desaturated product, arachidonic acid, for membrane structure, fluidity, skin integrity, and numerous tissue functions.

A critical function of dietary linoleic acid is to provide arachidonic acid, which is the precursor of the prostanoids and leukotrienes. These compounds modulate many vital physiological functions related to cardiovascular, renal, pulmonary, secretory, digestive, reproductive, and immune functions (Chapter 2). Recognition of the importance of arachidonic acid in the regulation of various cell functions under normal physiological and pathophysiological conditions has stimulated great interest in elucidating mechanisms and discovering specific inhibitors or agents that control the effects of arachidonic acid products [41]. The potential of dietary n-3 PUFAs in this regard is of great interest, as they may provide a dietary approach to control of arachidonic acid-mediated pathophysiological states.

The current minimum requirement for linoleic acid, or the amount needed to relieve or prevent symptoms of essential fatty acid deficiency, is 1.5% of dietary calories [37]. In recent years because of their ability to depress serum lipids, the consumption of PUFAs of vegetable oils has been strongly encouraged. Current consumption of linoleic acid ranges from 10 to 15 g/day in the United States. This high level of n-6 PUFA consumption could result in extremely high tissue levels of arachidonic acid, thereby promoting the synthesis of eicosanoids, which may exacerbate pathophysiological phenomena, i.e. thrombosis, arthritis, inflammation, immune functions, tumor growth, and psoriasis [38].

Metabolic imbalances resulting from excessive intake of specific nutrients is not uncommon, yet relatively little is known about the potential deleterious effects of consuming high levels of linoleic acid. Few nutrients are converted to such a range of bioactive agents as is this particular fatty acid.

Increased n-6 PUFA consumption raises questions about the optimum intake of n-6 PUFAs and whether intake levels should be based exclusively on serum cholesterol or reflect consideration of general health. Systematic research is needed to resolve these very important questions.

The possible essentiality of n-3 PUFAs has been discussed in detail by Tinoco [43] and more recently, with additional evidence, by Neuringer and Connor [44]. Based on their inability to relieve all symptoms of essential fatty acid deficiency, n-3 PUFAs have not been considered essential dietary components. However, based on the findings that n-3 PUFAs are essential components of nerve tissue, that they occur in high concentrations in sperm and in the outer rod segment of the eye, and that rigorous exclusion of n-3 PUFAs from the diet causes visual impairment, some now feel that even trace amounts of n-3 PUFAs are important in the diet because humans lack the enzymes to synthesize these fatty acids.

Neuringer et al. [44,45] suggested that n-3 PUFAs may be involved in visual acuity and learning ability. They also emphasized the importance of a dietary supply of n-3 PUFAs for the developing fetus and neonate, especially during the last trimester of pregnancy and the first three months of life when membranes containing high concentrations of n-3 PUFAs are being actively assembled.

In addition to their possible direct role in membrane structure and tissue functions, n-3 PUFAs such as linolenic, eicosapentaenoic, and docosahexaenoic acids may be important in modulating the metabolic interconversions of fatty acids of the n-6 series. In this regard, dietary n-3 PUFAs may moderate the production of eicosanoids in response to normal physiological stimuli and thereby prevent the pathophysiological states caused by the excessive production of eicosanoids derived from n-6 PUFAs. In addition, the n-3 PUFAs, particularly EPA, may be converted to eicosanoid analogs with potencies different from those derived from the n-6 PUFAs and thereby alter certain physiological responses.

Dietary linolenic acid is absorbed and metabolized to EPA and DHA to a limited extent in humans. The ability of dietary linolenic acid to reduce blood pressure suggests that dietary manipulation to incorporate n-3 PUFAs may be useful in the treatment and prevention of hypertension [42].

The actual and potential effects of dietary n-3 PUFAs found in fish oils and the manner in which they affect the metabolism of n-6

PUFAs needs careful, systematic research. The nutritional and physiological implications of dietary fish oil was briefly reviewed by Sanders [40]. It is possible that the apparent beneficial effects of dietary fish oil may reflect the modulation of the metabolism of dietary linoleic acid and arachidonic acid in response to specific stimuli. The potential effects of n-3 PUFAs as prophylactic or therapeutic agents for the reduction of serum lipids and thrombosis must be considered.

RECENT TRENDS IN MORTALITY DUE TO CARDIOVASCULAR DISEASE AND THEIR POSSIBLE CAUSES

Since 1950, the age-adjusted mortality rate from cardiovascular disease for the U.S. population has decreased by 40% [1]. Most of this reduction has taken place since 1968 [1]. Cardiovascular disease morbidity data are scanty, so it is difficult to determine whether the decline is due to the development of more effective treatments for disease victims or to a decrease in the incidence of disease. Some evidence links the decrease in mortality to the existence of better medical services [1]. National surveys indicate increased attention to reduction of risk, including the adoption of beneficial food habits [1,21]. Since the 1950s, the purchase of milk, cream, butter, eggs, and other foods with a relatively high cholesterol and saturated fat content has declined, and the use of vegetable oils rich in polyunsaturated fatty acids has increased by 75% [21]. During the last twenty years, the mean intake of cholesterol has dropped from 800 mg to less than 500 mg/day [22]. In the same period, mean plasma cholesterol has dropped 5% [21]. It has been proposed that of the 26% decrease in death due to coronary heart disease observed in the United States between 1968 and 1980, 10% or more could be attributed to the change in plasma cholesterol levels [21], some of which may be related to dietary unsaturated fatty acids.

It is probable that both the avoidance of risk factors and improved treatment methods have contributed to the reduction in deaths from cardiovascular disease. This decline, which has been attributed to alteration of dietary fat intake, also may be attributed to increased consumption of polyunsaturated fatty acids, which exert beneficial effects by lowering serum lipid and cholesterol levels, increasing membrane fluidity, and reducing thrombosis by influencing the conversion of arachidonic acid to eicosanoids. However, the mortality decrease is not a worldwide phenomenon, occuring only in countries where beneficial behavioral changes have been adopted. This fact supports the significance of the modification of risk factors, particularly dietary habits [22]. The exact mechanism underlying the effects of n-6 PUFAs were not understood until the prostaglandin pathway was elucidated and the roles of thromboxane and prostacyclin were defined.

STUDIES ON GREENLAND ESKIMOS

In the time since high polyunsaturated fat intake was recommended to prevent cardiovascular disease, n-6 PUFAs have been the fats most commonly consumed, particularly because they are readily available as major constituents of vegetable oils and their products. The n-3 PUFAs of marine oils were largely ignored until their potential positive effects were indicated by epidemiological studies carried out in the 1970s, which revealed a very low incidence of ischemic heart disease among Greenland Eskimos eating a traditional diet rich in fish and other seafoods. Compared to their Danish counterparts, Greenland Eskimos had lower levels of serum cholesterol, triglycerides, LDLs, and VLDLs and a higher level of HDLs, all characteristics related to a reduced risk of atherosclerosis (Figure 1.3). The Eskimos' protection from ischemic heart disease and associated illnesses was more strongly related to their lifestyle than to their genetic makeup. When Eskimos emigrated to Denmark and adopted a more western way of life, their serum cholesterol and lipid concentrations approached the normal Danish levels and their incidence of cardiovascular disease increased to levels typical for Europeans [24,25,26,27,36].

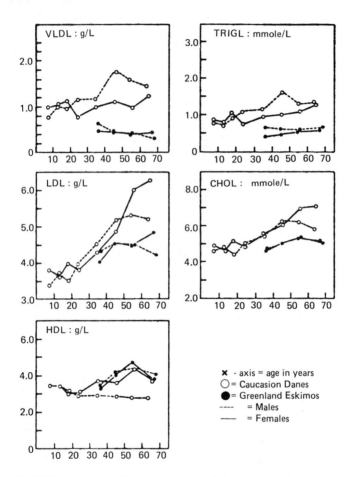

FIGURE 1.3 Mean plasma lipid and lipoprotein concentration in Greenland Eskimos and a Danish control group. (From Ref. 2; reprinted by permission from Blackwell Scientific Publications, Ltd.)

Because of their consumption of fish, seal, and whale, rich sources of n-3 PUFAs, the Eskimos' intake of these fatty acids is significantly higher than that of their Danish counterparts. Additionally, they consume less n-6 PUFAs (Table 1.2). The distribution of fatty acids in the serum and platelet lipids of Eskimos reflected the dietary lipids. In contrast to the Danish population, the platelet content of n-6 fatty acids, such as arachidonic acid

TABLE 1.1 Fatty Acid Composition of Plasma Phospholipids and Platelet Lipids in Greenland Eskimos and Danish Controls

Fatty acid	Plasma phospholipid (%)		Platelet lipids(%)	
	Eskimos	Danes	Eskimos	Danes
8:0-15:0	4.2	0.3	---	---
12:0-14:0	---	---	0.2	0.1
16:0	32.2	30.0	} 23.8	20.4
16:1	4.3	0.8		
16:2-17:1	---	0.7	---	---
18:0	17.0	14.6	12.0	17.2
18:1	14.9	12.4	18.2	17.2
18:2	5.4	22.3	4.0	8.2
18:3	0.1	0.2	---	---
18:4-20:3	3.3	0.7	---	---
20:0-20:1	---	---	5.6	2.3
20:4	3.8	7.4	8.9	22.1
20:5	7.4	1.8	8.3	0.5
22:0-22:5	2.2	2.2	---	---
22:0-24:1	---	---	8.8	5.9
22:5	---	---	3.4	1.0
22:6	3.6	2.2	6.1	1.5
24:0-24:1	2.5	4.2	---	---
20:5/20:4	1.95	0.24	0.93	0.02

Source: Ref. 24

TABLE 1.2 Dietary Fats in Eskimo and Danish Food Computed on a Daily Energy Consumption of 3,000 Kcal

Dietary fats		Eskimos	Danes
Fat energy	%	39	42
Saturated:	12:0	1.1	5.9
	14:0	3.7	7.5
	16:0	13.6	25.5
	18:0	4.0	9.5
	20:0	0.1	4.3
	% Total Fats	23	53
Monoenes:	16:1	9.8	3.8
	18:1	24.6	29.2
	20:1	14.7	0.4
	22:1	8.0	1.2
	% Total Fats	58	34
Polyenes:	18:2 n-6	5.0	10.0
	18:3 n-3	0.6	2.0
	20:5 n-3	4.6	0.5
	22:5 n-3	2.6	0
	22:6 n-3	5.9	0.3
	% Total Fats	19	13
P/S Ratio:		0.84	0.24
n-3 PUFAs (g/day):		14	3
n-6 PUFAs (g/day):		5	10
Cholesterol (g/day):		0.79	0.42

Source: Ref. 36

(AA) and linoleic acid (LA), were low and those of n-3 fatty acids, such as eicosapentaenoic acid (EPA) and docosahexaenoic acid (DHA), were severalfold greater (Table 1.1) [24,25,26,27,36]. The evidence that a diet rich in n-3 PUFAs is involved in the prevention of atherosclerosis and accounts for the low incidence of coronary heart disease in Eskimo populations was comprehensively reviewed by Dyerberg [46].

Platelet function (e.g. the tendency toward aggregation) is greatly reduced in plasma from Eskimos. A tendency for prolonged bleeding was noted, and in vitro experiments showed that platelet aggregation induced by adenosine diphosphate (ADP) or collagen was inhibited while aggregation induced by AA was not.

This behavior was related to alterations of the synthesis of prostaglandins from AA in platelets and vessel wall endothelium.
The two AA metabolites, thromboxane (TXA_2) and prostacyclin (PGI_2), exert opposite effects on platelet aggregation. It was hypothesized that alterations in n-3 and n-6 PUFA levels such as those observed in Eskimos resulted in altered TXA_2 and PGI_2 synthesis levels, causing reduced platelet aggregation and prolonged bleeding [24,25,26,27,28,36] (Chapter 2). It also was suggested that a prostaglandin which inhibits platelet aggregation (PGI_3) and another with only a weak platelet aggregating tendency (TXA_3) may be formed from EPA [29]. This was subsequently confirmed [34,35].

Thus, the low incidence of ischemic heart disease among Eskimos was originally hypothesized to be attributed to an increase in the proportion of n-3 PUFAs in dietary lipids; this led to beneficial (i.e lower) serum lipid levels and modifications of fatty acid composition of serum and platelet lipids, which reduced platelet aggregability. An increase in the plasmatic concentration of antithrombin III and antithrombin III activity and a higher frequency of a genetic variant in the complement system were also considered to be involved in the reduced incidence of disease. However, these factors associated with Eskimo genetic characteristics were considered of lesser significance [25] (Figure 1.4).

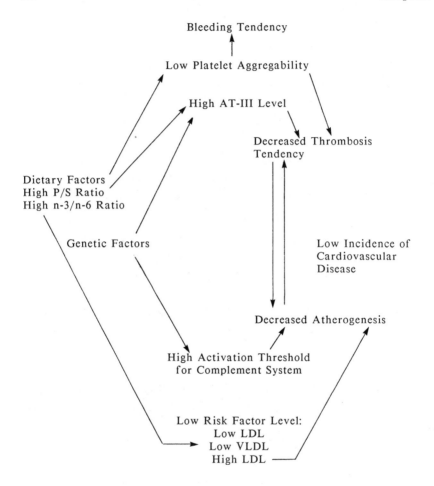

P/S: polyunsaturated/saturated fatty acids; n-3/n-6: n-3 and n-6 PUFAs; AT-III: antithrombin III.

FIGURE 1.4 Proposed mechanisms showing interaction of dietary and genetic factors leading to low incidence of cardiovascular disease among traditional Greenland Eskimos. (From Ref. 25.)

STUDIES ON OTHER POPULATIONS

Other investigations carried out in communities dissimilar to those of traditional Eskimos also reported associations between frequent fish consumption and a decreased risk of cardiovascular disease. In an epidemiological study of over 800 Welsh women, multiple regression analyses showed that the consumption of fatty fish had a strong positive association with HDL cholesterol [30].

Plasma n-3 PUFA levels were higher and platelet aggregability was lower among the residents of Japanese fishing villages than among those living in areas dedicated to farming [31]. Extensive epidemiological data show that Japanese fishermen who consume relatively large quantities of fish and seafoods have a lower incidence of coronary heart disease and atherosclerosis compared to Japanese farmers [47,48,49]. Fishermen consume from 150 to more than 200 g fish/day compared to farmers who consume 90 to 100 g/day.

In a group of more than 200 American coronary artery disease victims who were observed for periods of sixteen to nineteen years, those who followed a treatment that included a diet rich in fatty fish experienced a significantly longer survival period than the control group [32]. Finally, it was observed that fish consumption among 852 middle-aged Dutch men was inversely related to coronary disease mortality rates during the succeeding twenty years [33].

The following chapters explore the relationship between the development of cardiovascular disease and the consumption of polyunsaturated fatty acids of marine origin. The role of the fatty acids as precursors of prostaglandins, which also alter platelet and vessel wall characteristics, is outlined. Laboratory animal and human volunteer feeding trials with results supporting the epidemiological studies are discussed. Finally, some outstanding questions concerning the mechanisms and potential dietary uses of fish oils in the treatment of cardiovascular disease and other diseases such as cancer are reviewed.

REFERENCES

1. Feinleib, M. (1984). The magnitude and nature of the decrease in coronary heart disease mortality rate. Am. J. Cardiol. 54:2C-6C.

2. Levy, R.I. (1984). Introduction. Am. J. Cardiol. 54:1C.

3. Mancini, M., and P. Rubba. (1983). Hyperlipidemia as a risk factor. In Arterial Pollution, H. Peeters, G.A. Gresham, R. Paoletti (eds). p. 271-281. Plenum Press, New York.

4. Johansen, K. (1982). Aneurysms. Scientific American 247(1):110-118.

5. Gresham, G.A. (1983). Atherosclerosis: its origin and development in man. In Arterial Pollution, H. Peeters, G.A. Gresham, R. Paoletti (eds). p. 271-281. Plenum Press, New York.

6. Ross, R. and J. Glomset. (1976). The pathogenesis of atherosclerosis. N. Eng. J. Med. 295:369-377.

7. Gresham, G.A. (1983). Atherosclerosis: some components of the lesion. In Arterial Pollution. H. Peeters, G.A. Gresham, R. Paoletti (eds). p. 79-90. Plenum Press, New York.

8. Walton, K.W. (1983). Functional aspects of atherosclerosis. In Arterial Pollution, H. Peeters, G.A. Gresham, R. Paoletti (eds). p. 23-53. Plenum Press, New York.

9. Grundy, S. M. (1979). Dietary fats and sterols. In Nutrition, Lipids, and Coronary Heart Disease. R.I. Levy, B.M. Rifkind, B.H. Dennis, N. Ernst (eds). p. 89-118. Raven Press, New York.

10. Herbert, P.N., P.D. Thompson, and R.S. Shulman. (1983). Determinants of plasma high density lipoprotein concentration. In Arterial Pollution. H. Peeters, G. A. Gresham, R. Paoletti (eds). p. 183-211. Plenum Press, New York.

11. Smith, E.B. (1982). Metabolism of the arterial wall. In Factors in the Formation and Regression of the Atherosclerotic Plaque, G.V.R. Born, A.L. Catapano, R. Paoletti (eds). p. 21-33. Plenum Press, New York.

12. Ross R., E. Raines, D. Bowen-Pope. (1982). Growth factors from platelets, monocytes, and endothelium: their role in cell proliferation. Ann. N.Y. Acad. Sci. 397:18-24.

13. Ross, R., D. Bowen-Pope, E.W. Raines, A. Faggiotto. (1982). Endothelial injury: blood-vessel wall interactions. Ann. N.Y. Acad. Sci. 401:260-264.

14. Born, G.V.R., Kratzer, M.A.A. (1982). Endogenous factors in platelet thrombosis. In <u>Factors in the Formation and Regression of the Atherosclerotic Plaque</u>, G.V.R. Born, A.L. Catapano, R. Paoletti (eds). p. 197-207. Plenum Press, New York.

15. Castelli, W.P. (1984). Epidemiology of coronary heart disease: the Framingham study. <u>Am. J. Med.</u> 76(2A):4-12.

16. Gresham, G.A. (1983). Atherosclerosis: clinical features, prevention, and regression. In <u>Arterial Pollution,</u> H. Peeters, G.A. Gresham, R. Paoletti (eds). p. 295 313. Plenum Press, New York.

17. Goodnight, S.H. W.S. Harris, W.E. Connor, D.R. Illingworth, (1982). Polyunsaturated fatty acids, hyperlipidemia and thrombosis. <u>Arteriosclerosis</u> 2(2):87-113.

18. Gotto, A.M., J. Shepherd, L.W. Scott, E. Manis. (1979). Primary hyperlipidemis and dietary management. In <u>Nutrition, Lipids, and Coronary Heart Disease</u>, R.I. Levy, B.M. Rifkind, B.H. Dennis, N. Ernst (eds). p. 247-283. Raven Press, New York.

19. Sirtori, C.A. (1982). Plasma lipoprotein changes induced by diets and drugs. In <u>Factors in Formation and Regression of the Atherosclerotic Plaque,</u> G.V.R. Born, A.L. Catapano, R. Paoletti (eds). p. 163-195. Plenum Press, New York.

20. Jorgensen, K.A. and J. Dyerberg. (1983). Platelets and atherosclerosis. <u>Adv. Nutr. Res.</u> 5:57-75.

21. Brown, W.V., H. Ginsberg, and W. Karmally. (1984). Diet and the decrease in coronary heart disease. <u>Am. J. Cardiol.</u> 54:27C-29C.

22. Levy, R.I. (1984). Causes of the decrease in cardiovascular mortality. <u>Am. J. Cardiol.</u> 54:7C-13C.

23. Ackman, R.G. (1980). Fish lipids. Part I. In <u>Advances in Fish Science and Technology,</u> pp. 86-103. J.J. Connell (ed) Fishing News Books, Ltd. Farnham, Surrey, England.

24. Dyerberg, J. (1981). Platelet-vessel wall interaction: influence of diet. <u>Philos. Trans. R. Soc. Lond.</u> B294:373-381.

25. Dyerberg, J. and H.O. Bang. (1982). A hypothesis on the development of acute myocardial infarction in Greenlanders. <u>Scand. J. Clin. Lab. Invest.</u> 42 (Suppl. 161):7-13.

26. Dyerberg, J. and K.A. Jorgensen. (1982). Marine oils and thrombogenesis. <u>Prog. Lipid Res.</u> 21:255-269.

27. Dyerberg, J. (1982). Observations in populations in Greenland and Denmark. In *Nutritional Evaluation of Long Chain Fatty Acids in Fish Oil*, S. M. Barlow and M.E. Stansby, (eds) p. 245--266. Academic Press, London.

28. Sanders, T.A. (1983). Dietary fat and platelet function. *Clin. Sci.* 65:343-350.

29. Jorgensen, K.A. and J. Dyerberg. (1983). Platelets and athero sclerosis. *Adv. Nutr. Res.* 5:57-75.

30. Yarnell, J.W.G., J. Milbank, C.L. Walker, A.M. Fehily and T.M. Hayes. (1982). Determinants of high density lipoprotein and total cholesterol in women. *J. Epid. Comm. Health* 36:167-171.

31. Hirai, A., T. Hamazaki, T. Terano, T. Nishikawa, Y. Tamura and A. Kumagai. (1980). Eicosapentaenoic acid and platelet function in Japanese. *Lancet* ii:1132-1133.

32. Nelson, A. (1972). Diet therapy in coronary disease: effect on mortality of high-protein, high seafood, fat-controlled diet. *Geriatrics* 27(12):103-116.

33. Kromhout, D. E.B. Bosschieter, C. de L. Coulander. (1985). The inverse relation between fish consumption and 20-year mortality from coronary heart disease. *N. Eng. J. Med.* 312(19):1205-1209.

34. Fischer, S. and Weber, P.C. (1984). Prostaglandin I3 is formed in man after dietary EPA. *Nature* 307:165-68.

35. von Schacky, C., S. Fischer, and P.C. Weber. (1985). Long-term effects of dietary marine w-3 fatty acids upon plasma and cellular lipids, platelet function, and eicosanoid formation in humans. *J. Clin. Invest.* 76:1626-31.

36. Dyerberg, J. (1986). Linolenate-derived polyunsaturated fatty acids and prevention of atherosclerosis. *Nutrition Reviews* 44(4):125-134.

37. Holman, R. and R. Kunau. (1977). *Polyunsaturated Fatty Acids*. American Oil Chemists Society. Champagne, Illinois.

38. Lands, W.M. (1986). *Fish and Human Health*. Academic Press, New York.

39. Ross, R. (1986). The pathogenesis of atherosclerosis: an update. *N. Eng. J. Med.* 314:488-500.

40. Sanders, T.A.B. (1986). Nutritional and physiological implications of fish oils. *J. Nutr.* 116:1857-60.

41. Samuelsson, B., F. Berti, G.C. Folco, and G.P. Velo (1985). <u>Drugs Affecting Leukotrienes and Other Eicosanoid Pathways.</u> Plenum Press, New York.

42. Berry, E.M. and J. Hirsch. (1986). Does dietary linolenic acid influence blood pressure? <u>Am. J. Clin. Nutr.</u> 44:336-40.

43. Tinoco, J. (1982). Dietary requirement and functions of alpha-linolenic acid in animals. <u>Prog. Lipid Res.</u> 21:1-45.

44. Neuringer, M. and W.E. Connor (1986). N-3 fatty acids in the brain and retina: evidence for their essentiality. <u>Nutr. Rev.</u> 44:285-294.

45. Neuringer, M., W.E. Connor, D.S. Lin, L. Barstad, and F. Luck (1986). Biochemical and functional effects of prenatal and postnatal n-3 fatty acid deficiency on retina and brain in rhesus monkeys. <u>Proc. Natl. Acad. Sci. USA.</u> 83:4021-25.

46. Dyerberg, J. (1986). Linoleinate derived polyunsaturated fatty acids and prevention of atherosclerosis. <u>Nutr. Rev.</u> 4:125-34.

47. Kagawa, Y., M. Nishizawa, M. Suzuki, T. Miyatake, T. Hamamoto, K. Goto, E. Motonaga, H. Izumikawa, H. Hirata, and A. Ebihara (1982). Eicosapolyenoic acids of serum lipids of Japanese islands with low incidence of cardiovascular disease. <u>J. Nutr. Sci. Vitaminol.</u> 28:441-53.

48. Kimura, N. (1983). Changing patterns of coronary heart disease, stroke, and nutrient intake in Japan. <u>Prev. Med.</u> 12:222-27.

49. Yamori, Y., Y. Narai, N. Iratani, R.J. Workman, and T. Inajami (1985). Comparison of serum phospholipid fatty acids among fishing and farming Japanese populations and American islanders. <u>J. Nutr. Sci. Vitaminol.</u> 31:417-22.

2
Eicosanoid Metabolism: General Background and Overview

INTRODUCTION

The beneficial effects of dietary polyunsaturated fatty acids on certain physiological functions have been attributed to their ability to lower serum lipids and cholesterol levels, to increase membrane fluidity, to depress triglyceride synthesis, and most recently to their metabolic role as precursors of eicosanoids. Linoleic acid, an essential fatty acid of the n-6 family, may be the most important precursor of the eicosanoids. It is sequentially desaturated and elongated by Δ-6 desaturase and elongase enzymes (Figure 1.2) to form arachidonic acid, from which the eicosanoids originate. Most of the early research examining the effects of dietary fat, particularly dietary linoleic acid, on prostaglandin synthesis was comprehensively reviewed by Hornstra [1] and Galli [1a].

The eicosanoids include prostaglandins and leukotrienes, a family of bioactive lipids produced mostly from arachidonic acid via cyclooxygenase and lipoxygenase, respectively. Cyclooxygenase produces the prostaglandin endoperoxides PGG_2 and PGH_2, which are precursors of prostacyclin (PGI_2), thromboxane (TXA_2), prostaglandin E_2 (PGE_2), and various other prostaglandin species that exert potent bioactive effects in specific tissues and/or organs. Lipoxygenase produces hydroperoxy fatty acids and leukotrienes, which are important agents in mediating immune functions and cellular interactions (Figure 2.1).

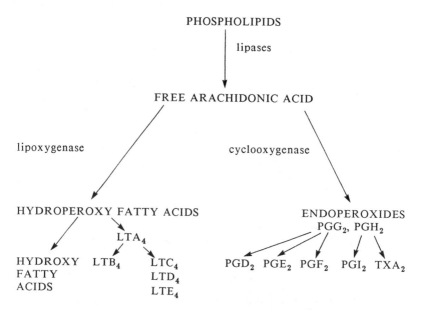

FIGURE 2.1 Lipoxygenase and cyclooxygenase pathways leading to the synthesis of eicosanoids.

Prostaglandins modulate many vital secretory, digestive, reproductive, immune, and circulatory functions [1-5]. The leukotrienes are involved in pulmonary and macrophage functions; allergic, immune, and inflammatory responses; and chemotaxis and chemokinesis [6,7]. The function and effects of the various prostaglandins and leukotrienes in blood and vascular cell functions, with emphasis on their homeostatic mechanisms, were discussed in depth by Gerrard [7a].

Eicosanoids are apparently produced in trace amounts (picomoles) in normal tissue. However, altered production of prostaglandins and/or leukotrienes has been associated with the development of a number of diseases, including atherosclerosis, thrombosis, inflammatory diseases, and asthma; eicosanoids also may be involved in the development of cancer and the growth and metastasis of tumors.

ARACHIDONIC ACID

Arachidonic acid can be consumed in the diet in small quantities; however, most of it is formed from dietary linoleic acid through desaturation and elongation reactions (Figure 2.2). Arachidonic acid is released from cellular phospholipids by phospholipases in response to particular stimuli that act mostly at the cell surface. Little information is available concerning the regulation of phospholipases or the manner in which they control the liberation of arachidonic acid. The quantity of arachidonic acid released reflects the amount in particular pools of tissue phospholipids. There is some concern that with the high consumption of polyunsaturated fatty acids in the American diet, tissue

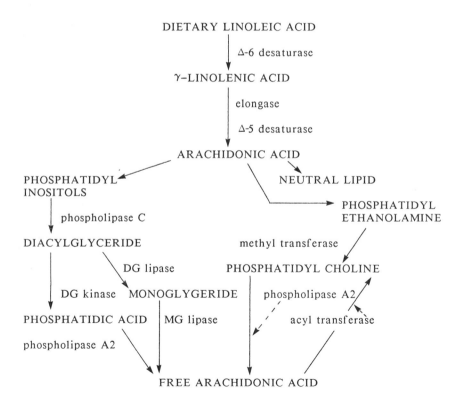

FIGURE 2.2 Release of arachidonic acid from cellular phospholipases.

phospholipid pools may be saturated with arachidonic acid. Hence, an excess may be released in response to normal physiological stimuli, particularly in diseased and/or stressed states, resulting in overproduction of eicosanoids [8].

Dietary linolenic acid is absorbed and metabolized to EPA and DHA to a limited extent in humans. Adam et al. [8a] showed that when 16% of dietary fat was obtained from this source, linoleic acid provided 4 en%. Although there was no appreciable depression of total arachidonic acid in lipid pools, excretion of prostaglandin metabolites decreased from 52% to 85%. There apparently may have been selective depression of arachidonic acid in certain tissue phospholipid pools and/or inhibition by n-3 PUFAs of PG synthetase.

EICOSANOID SYNTHESIS

Most mammalian tissues are capable of synthesizing certain types of eicosanoids, depending on the appropriate series of enzymes they possess. There are marked differences between species in the presence of cyclooxygenase and lipoxygenase enzymes; hence, one has to be careful in extrapolating observations from animals to humans [27,28]. The type of eicosanoid synthesized varies from tissue to tissue and from cell to cell. PGE_2 is produced by neutrophils, macrophages, and platelets and is the predominant product in the kidney and seminal vesicle gland. TXA_2 is the major metabolite in platelets, and PGI_2 is the sole prostanoid produced by the arterial endothelial cells [9].

The involvement of neutrophils and macrophage cells in the thrombotic process is now recognized. Neutrophils serve as a source of leukotrienes, e.g., LTB_4, which induce neutrophil aggregation, adhesion to endothelial cells, chemotaxis, and plasma exudation. These effects may cause local inflammation and aid atherogenesis [31].

Prostaglandins

Arachidonic acid is the precursor of prostaglandins of the 2-series, such as TXA_2, PGE_2, and PGI_2. In platelets the endoperoxides are transformed into a 17-carbon 12-hydroxy fatty acid (HHT),

Eicosanoid Metabolism

malondialdehyde (MDA), and TXB_2 via TXA_2, which is very unstable with a half-life of about 30 seconds. TXB_2 is its stable metabolite (Figure 2.3). TXA_2 has very strong platelet aggregating and vasoconstrictive effects.

Various eicosanoid metabolites of arachidonic acid exert profound modulatory effects on cells involved in immune-inflammatory responses and in immune regulation. The pervasive role of eicosanoids in the regulation and/or modulation of monocyte, leukocyte, macrophage, and neutrophil functions in relation to inflammation was recently summarized by Kunkel and Chensue [32]. Macrophages, for example, are copious producers of PGE_2, leading to local vasodilation and edema (i.e., they are proinflammatory); but in addition, PGE_2 dramatically suppresses the activity of macrophage-lymphocyte-mediated reactions, including B and T cell proliferation, natural killer cell activity, and cellular-mediated cytotoxicity. Furthermore, there is increasing evidence that PGE_2 may act as an immunosuppressor and facilitate tumor growth [10,32]. The ability of dietary n-3 PUFAs to depress the production of eicosanoids by

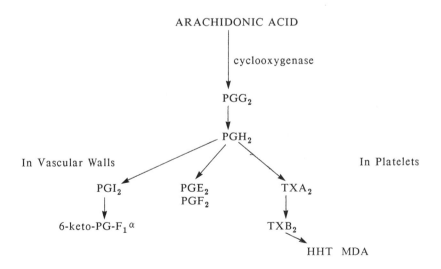

FIGURE 2.3 Synthesis of prostaglandins.

TABLE 2.1 Summary of Prostaglandin Involvement in Physiological Processes and Pathological States

---

* Glucose homeostasis and diabetes mellitus
* Hypersensitivity and inflammation
* Renal function, vascular regulation, and hypertension
* Reproductive processes
* Hemostasis, thrombosis, and thromboembolic disorders
* Pulmonary disease
* Gastrointestinal secretion and motility
* Peptic ulcer disease
* Host defense in cancer
* Hypercalcemia of cancer

---

macrophages may partly explain the beneficial effects of these fatty acids on tumor growth.

In the endothelial cells lining the vascular walls, endoperoxides are converted to PGI_2, which has a half-life of about 3 minutes and decomposes into the more stable 6-keto PGF_1. PGI_2 is a potent inhibitor of platelet aggregation as well as a vasodilator (Table 2.1).

The antagonistic effects of TXA_2 and PGI_2 are important in modulating hemodynamics and cardiovascular function. For example, platelet aggregation is controlled by TXA_2 production from platelet arachidonic acid. Excessive TXA_2 causes vasoconstriction and platelet clumping, which may lead to vascular thrombosis. Platelet aggregation and vasoconstriction is counteracted by the vasodilating effects of PGI_2 produced from vascular tissue arachidonic acid [1,2,23] (Table 2.2).

TABLE 2.2 Biological Functions of Prostanoids

Organ system	Effects	Active species
Blood vessels	vasodilation	$PGI_2 > PGI_3 > PGE_1$
	vasoconstriction	TXA_2
Platelets	adhesion, aggregation	TXA_2
	antiaggregatory	$PGI_2 > PGI_3 > PGE$
Lung	bronchiole constriction	PGF_2, TXA_2, PGD_2
	bronchiole dilation	PGE_2, PGI_2
Kidney	g. filtration rate	PGE_2, PGI_2, TXA_2
	renin secretion	PGI_2, PGF_2
	naturesis	PGE_2, PGI_2
	diuresis	PGE_2
Stomach	acid secretion	PGE_2, PGE_1
Small intestine	peristalsis	PGE_2, PGF_2
Pancreas	secretion	PGE_2, PGI_2
	amylase secretion	PGE_2, PGI_2
	insulin secretion	PGE
Hypophysis	secretion GH, ACTH	PGE, PHF
Tissue	pain	PGE_2, PGE_2
	cytoprotection	PGI_2, dimethyl PGE

The polyunsaturated fatty acids of fish oils, such as eicosapentaenoic acid (EPA) and docosahexaenoic acid (DHA), may alter prostaglandin synthesis by effectively competing with arachidonic acid for cyclooxygenase. Although EPA is normally a poor substrate for the enzyme, under high peroxidative conditions (high peroxide tone) EPA may be converted to PGI_3, which inhibits platelet aggregation. EPA may also be converted to small quantities of TXA_3, a prostaglandin with only weak aggregation-inducing activity, per se, but which binds to the TXA_2 binding site, thereby reducing the effectiveness of this aggregating agent [24,25,26].

Normally prostaglandins are formed in short-lived bursts by many tissues in response to specific physiological stimuli that vary with the tissue. Prostaglandins can act on these cells or receptors on the surface of neighboring cells to elicit responses. Although prostaglandins are rapidly inactivated and have half-lives of only one to three minutes, they may have systemic effects when produced in large quantities or continuously in pathological states and thereby induce several metabolic perturbations [11].

Leukotrienes

Leukotrienes are acyclic, oxygenated eicosanoids containing three alternating conjugated double bonds. They are produced primarily from tissue arachidonic acid by the action of lipoxygenases [7]. Leukotrienes are produced by several types of cells, including platelets, lymphocytes, neutrophils, mast cells, and macrophages, as well as tissues of the lung, coronary and pulmonary arteries, skin, and eye [12]. Leukotrienes of the 4-series--A_4, B_4, C_4, D_4, E_4, and F_4--are generated via the 5-lipoxygenase pathway, although in some tissues these may be generated by 12- and 15-lipoxygenases [7]. Leukotriene A_4, an unstable intermediate, is converted enzymatically to leukotriene B_4 and, by the addition of glutathione, to leukotriene C_4. Leukotrienes D_4, E_4, and F_4 are metabolites of C_4, and all three form the family of cysteinyl-containing leukotrienes [13] (Figure 2.4). The 5-series of leukotrienes, which are less biologically active [14], can be derived from EPA [6].

Leukotrienes exhibit a number of potent biological effects: vasoconstriction and bronchoconstriction; enhancement of vascular permeability; erythema; and attraction and activation (chemotaxis and chemokinesis) of white blood cells. In addition, they affect mucus secretion in the respiratory tract, cause vascular dilation, and exert a negative inotropic effect on cardiac contraction [7,12,13].

Leukotrienes are important in many tissue inflammatory and hypersensitivity reactions. They cause adhesion of leukocytes to the endothelium, stimulate chemotaxis of neutrophils, and may enhance prostaglandin synthesis by macrophages [13] (Table 2.3).

Eicosanoid Metabolism

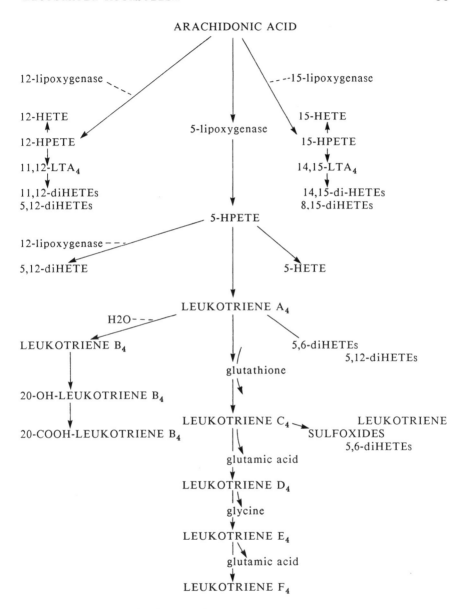

FIGURE 2.4 Synthesis of leukotrienes.

TABLE 2.3 Biological Actions of Leukotrienes

Organ System	Effects	Species
Bronchioles	constriction	LTC_4, LTD_4
Ileum	constriction	LTC_4, LTD_4
Vascular	constriction permeability	LTC_4, LTD_4 LTC_4, LTD_4
Pancreas	insulin secretion	LTB_4, HETE
Neutrophils	adhesion chemotaxis/-kinesis lysozyme secretion	LTB_4 LTB_4, HETE(5,9,11) LTB_4, HETE(5,12)
Monocytes	chemotaxis/-kinesis	LTB_4, HETE(5,9,11)
Basophils	histamine secretion	LTB_4, HETEs

The cysteinyl-containing leukotrienes constrict arterioles and simultaneously increase the permeability of postcapillary venules. There is some evidence that leukotrienes may play a role in cardiac anaphylaxis [7].

The potency of the different leukotrienes varies with type and animal species. In humans they play a central role in allergic bronchial constriction. Leukotrienes C_4 and D_4 stimulate mucus secretion in the trachea of various species. Leukotriene B_4 promotes leukocyte infiltration into the lungs. Evidence indicates that leukotrienes C_4, D_4, and E_4, in addition to histamine and some of the new neuropeptides, are important agents for modulating pulmonary functions and may be actively involved in bronchial asthma and anaphylaxis [7].

Much of the current research on the effects of leukotrienes has focused on pathophysiological phenomena, including inflammation, allergy, psoriasis, asthma, and cancer. However, leukotrienes are also important in normal physiological functions. Interaction between the prostaglandin and leukotriene pathways may occur in certain

Eicosanoid Metabolism

tissues; for example, PGE_2 inhibits the release of leukotriene B_4 from rat neutrophils [15] and both eicosanoids act in concert for the potentiation of inflammatory reactions [16].

In addition to their chemotactic and chemokinetic properties, leukotrienes apparently are involved in immunological mechanisms; for example, leukotriene B_4 affects lymphocyte activation and function and activates T-lympohcytes to become suppressor cells [7].

PUFAs and Leukotrienes

In view of the involvement of leukotrienes in various diseases, researchers are actively seeking agents that can control the synthesis of leukotrienes in various tissues. Because leukotrienes are synthesized from arachidonic acid, it is conceivable that fish oils can affect leukotriene metabolism. By inhibiting cyclooxygenase and reducing PGE_2 synthesis, the n-3 PUFAs of fish oils could divert arachidonic acid into the lipoxygenase pathway.

Only a limited number of studies have examined the effects of dietary fish and/or fish oils on leukotriene synthesis. In mice, dietary menhaden oil depressed arachidonic acid concentration and increased EPA concentration in mastocytoma cell phospholipids. Following activation, these cells decreased their production of leukotriene B and synthesized the less potent leukotrienes of the 5-series [20].

When the synthesis of leukotrienes in the peritoneal cavities of rats maintained on a menhaden oil supplemented diet was studied, tissue levels of EPA increased, and AA tissue levels decreased. Leukotrienes of the 5-series accounted for more than 30% of the total leukotrienes synthesized. Leukotriene C_5 was as potent as leukotriene C_4 in contracting smooth muscle. Leukotriene B_5 was less potent than leukotriene B_4 in eliciting neutrophil chemotaxis, indicating that leukotriene B_5 is a weak, partial agonist [20-22,29].

The conversion of EPA via lipoxygenase into 5-series leukotrienes occurs in many cells that are involved in the genesis of an inflammatory response in various diseased states, particularly neutrophils. Prescott [17] showed that cultured human neutrophils progressively esterify exogenous EPA into phospholipids mostly at the expense of arachidonic acid. Dietary fish oils significantly reduced the production of leukotriene B_4, thereby limiting its impact on inflammation.

In human feeding studies [18,19] dietary fish oils significantly increased the EPA content of neutrophils and monocytes and caused negligible changes in AA and DHA content. The production of leukotriene B_4 was reduced and the LTB_4-mediated functions of neutrophils were inhibited, suggesting a beneficial role of n-3 PUFAs in suppressing inflammatory responses.

EPA and DHA are competitive inhibitors of tissue lipoxygenase. Compared to arachidonic acid, EPA is a preferred substrate and DHA is a markedly inferior substrate for 5-lipoxygenase in human neutrophils [18]. In activated human neutrophils, DHA had little effect on arachidonic acid metabolism; EPA reduced synthesis of leukotriene B_4 and was converted to leukotriene B_5, reducing the chemotactic and aggregating activities of the neutrophils [21].

Dietary fish oils inhibit the release of arachidonic acid, reduce the synthesis of leukotrienes of the 4 series, affect the adherence of neutrophils, and reduce the chemotactic and aggregating activities of cells [17-22]. By inhibiting the 5-lipoxygenase pathway in neutrophils and monocytes and by reducing leukotriene B_4-mediated functions, diets enriched with fish oil may attenuate inflammatory responses [18]. This antiinflammatory effect may also reduce the development of arterial plaque as well as minimize inflammation in injured or stressed tissue. The suppression of inflammatory response may be particularly important in patients with gout, rheumatoid arthritis, psoriasis, degenerative joint disease, systemic lupus erythematosis, and other chronic inflammatory diseases [18,30].

CONCLUSION

Because of the potent and pervasive effects of eicosanoids as mediators of cardiovascular integrity and hemodynamics, hypersensitivity and inflammation, vascular permeability, smooth muscle contraction, and vasoconstriction (Tables 2.2 and 2.3), reducing the production of eicosanoids may reduce and/or ameliorate pathophysiological states that reflect excessive intake of n-6 PUFAs. Further studies and clinical trials with humans consuming fish or fish oils are urgently needed [13].

REFERENCES

1. Hornstra, G. (1982). Dietary Fats, Prostanoids and Arterial Thrombosis. Martinus Nijhoff Publishing, Boston.

1a. Galli, C. (1980). Dietary influences on prostaglandin synthesis. Adv. Nutr. Res. 3:95.

2. Marcus, A.J. (1984). The eicosanoids in biology and medicine. J. Lipid Res. 25:1511-16.

3. Jacobsen, D.C. (1983). Prostaglandins and cardiovascular disease; a review. Surgery 93:564-73.

4. Ninnemann, J.L. (1984). Prostaglandins and immunity. Immunol. Today 5:170.

5. Sanders, T.A.B. (1984). Dietary polyunsaturated fatty acids and health. Brit. J. Clin. Pract. (Symp. Suppl.)31:24-32.

6. Hammarstrom, S. (1983). Leukotrienes. Ann. Rev. Biochem. 52:35577.

7. Stjernschantz, J. (1984). The leukotrienes. Medical Biology 62:215-30.

7a. Gerrard, J.M. (1985). Prostaglandins and Leukotrienes. Marcel Dekker, New York.

8. Lands, W.E.M. (1985). Fish and Human Health. Academic Press, New York.

8a. Adam, O., G. Wolfram, and N. Zollner. (1986). Effect of alpha-linolenic acid in the human diet on linoleic acid metabolism and prostaglandin biosynthesis. J. Lipid Res. 27:421-26.

9. Kuehl, F.A. (1980). Prostaglandins, arachidonic acid, and inflammation. Science 210:978-84.

10. Goodwin, J.S. and J.L. Ceuppens. (1985). Prostaglandins, cellular immunity and cancer. In Prostaglandins and Immunity, pp.1-34. J.S. Goodwin (ed) Martinus Nijhoff Publishing, Boston.

11. Smith, W.L. (1985). Cellular and subcellular compartmentation of prostaglandin and thromboxane synthesis. In Biochemistry of Arachidonic Acid Metabolism, pp.77-93. W.E.M. Lands (ed) Martinus Nijhoff Publishing, Boston.

12. Ophir, J., S. Brenner, and S. Kivity. (1985). Leukotrienes. Int. J. Derm. 24:199-203.

13. Samuelsson, B. (1983). Leukotrienes: mediators of immediate hypersensitivity reactions and inflammation. Science 220:568-75.

14. Lee, T.H., K.F. Austen, E.J. Corey, and J.M. Drazen. (1984). LTE_4-induced airway hyperresponsiveness of guinea pig tracheal smooth muscle to histamine. Proc. Natnl. Acad. Sci. 81:4922-25.

15. Ham, E.A., D.D. Soderman, M.E. Zanetti, H.W. Dougherty, E. McCauley, and F.A. Kuehl. (1983). Inhibition by prostaglandins of leukotriene B_4 release from activated neutrophils. Proc. Natl. Acad. Sci. 80:4349-53.

16. Maclouf, J. (1985). Cell-cell signalling. In Biochemistry of Arachidonic Acid Metabolism, pp.298-309. W.E.M. Lands (ed) Martinus Nijhoff Publishing, Boston.

17. Prescott, S.M. (1984). The effect of eicosapentaenoic acid on leukotriene B production by human neutrophils. J. Biol. Chem. 259(12):7615-21.

18. Lee, T.H., R.L. Hoover, J.D. Williams, R.J. Sperling, J. Ravalese, B.W. Spur, D.R. Robinson, E.J. Corey, R.A. Lewis, and K.F. Austen. (1985). Effect of dietary enrichment with eicosapentaenoic and docosahexaenoic acids on in vitro neutrophil and monocyte leukotriene generation and neutrophil function. New Eng. J. Med. 312:1217-24.

19. Lee, T.H., J-M. Mencia-Huerta, C. Shih, E.J. Corey, R.A. Lewis, and K.F. Austen. (1984). Effects of exogenous arachidonic, eicosapentaenoic, and docosahexaenoic acids on the generation of 5-lipoxygenase pathway products by ionophore-activated human neutrophils. J. Clin. Invest. 74:1922-33.

20. Murphy, R.C., W.C. Pickett, B.R. Culp, and W.E.M. Lands. (1981). Tetraene and pentaene leukotrienes: selective production from murine mastocytoma cells after dietary manipulation. Prostaglandins 22:613-22.

21. Lee, T.H., J.M. Mencia-Huerta, C. Shih, E.J. Corey, R.A. Lewis, and K.F. Austen. (1984). Characterization and biologic properties of 5,12-dihydroxy derivatives of eicosapentaenoic acid, including leukotrienes B_5 and double lipoxygenase product. J. Biol. Chem. 259:2383-89.

22. Lee, T.H., J.M. Drazen, R.A. Lewis, and K.F. Austen. (1985). Substrate and regulatory functions of eicosapentaenoic and docosahexaenoic acids for the 5-lipoxygenase pathway. Implications for pulmonary responses. Prog. Biochem. Pharmacol. 20:1-17.

23. Levin, R.I., B.B. Weksler, A.J. Marcus, and E.A. Jaffe. (1984). Prostacyclin production by endothelial cells. In Biology of Endothelial Cells, p 288. E.A. Jaffe (ed) Martinus Nijhoff Publishing, Boston.

24. Dyerberg, J. (1981). Platelet vessel wall interaction: influence of diet. Philos. Trans. Royal Soc. London B294:373-381.

25. Fischer, S. and P.C. Weber. (1984). Prostaglandin I_3 is formed in man after dietary EPA. Nature 307:165-68.

26. Von Schacky, C., S. Fischer, and P.C. Weber. (1985). Long-term effects of dietary marine omega-3 fatty acids upon plasma and cellular lipids, platelet function, and eicosanoid formation in humans. J. Clin. Invest. 76:1626-1631.

27. Bruckner, G.G., B. Lokesh, B. German, and J.E. Kinsella. (1984). Biosynthesis of prostanoids, tissue fatty acid composition, and thrombotic parameters in rats fed diets enriched with docosahexaenoic or eicosapentaenoic acids. Thromb. Res. 34:479-97.

28. Morita, I., R. Takahashi, Y. Saito, and S. Murota. (1983). Effects of eicosapentaenoic acid on arachidonic acid metabolism in cultured vascular cells and platelets. Species difference. Thromb. Res. 31(2):211-17.

29. Leitch, A.G., T.H. Lee, E.W. Ringel, J.D. Prickett, D.R. Robinson, S.G. Pyne, E.J. Corey, J.M. Drazen, K.F. Austin, and R.A. Lewis. (1984). Immunologically induced generation of tetraene and pentaene leukotrienes in the peritoneal cavities of menhaden-fed rats. J. Immunology 132:2559-64.

30. Editors. (1986). Dietary fish oil alters leukotriene generation and neutrophil function. Nutrition Reviews. 44(4): 137-39.

31. Marcus, A.J., R.R. Gorman, and J.W. Ward (1985). Inhibition of platelet function in thrombosis. Circulation 72:698-702.

32. Kunkel, S.L. and S.W. Chensue (1986). The role of arachidonic acid metabolites in mononuclear phagocytic cell interactions. Int. J. Dermatol. 25:83-91.

3
The Effects of Omega-3 Polyunsaturated Fatty Acid Consumption on the Plasma, Platelet, Vessel Wall, and Erythrocyte Characteristics of Human Subjects in Feeding Trials

INTRODUCTION

Epidemiological observations of the beneficial effects of n-3 PUFAs have prompted numerous human feeding trials. Volunteers have consumed fatty fish [1-12,47], fish oils [13-21,53], fish oil concentrates [22-42,49] or purified EPA [43-46,48,55] to confirm and obtain more specific information about the mechanisms of the ascribed beneficial effects of n-3 PUFAs on cardiovascular functions. These studies are reviewed and the findings are compared to the plasmatic and platelet characteristics of Eskimos on prolonged fish oil diets. The results are summarized in Tables 3.1 to 3.5.

Feeding Trials: Sources of Dietary N-3 PUFAs

Diets rich in fatty fish such as mackerel, salmon, and herring were fed for periods ranging from one week [2] to three months [3,5]. Fish was consumed in quantities ranging from that providing all or nearly all the subjects' protein and fat intake [1,3,4,5,7,9] to that served at two weekly meals while maintaining the participants' usual diets at other times [6] (Table 3.1).

Salmon oil was consumed as a component of a diet abundant in salmon meat [4,5,9,10,11]. The daily amounts ranged from 60 to 90 ml (or from three to six tablespoons) for a period of four weeks. When cod liver oil was used as a dietary supplement, the participants maintained their usual diets. Cod liver oil was consumed in amounts varying between 6.8 g [13] and 40 ml/day [8,20,21,53] for periods of two [13,14,18], six [15,16,17,19] and 20 weeks [53]. The quantity of

TABLE 3.1 Human Feeding Trials: Experimental Design

Reference	Experimental design
	I. FATTY FISH
1	Crossover study; 2 3-week periods of either control or fish diet Subjects: 42 19♂ 24-76 years 23♀ 30-53 years Fish: mackerel Dose: 200 g/day Feeding period: 3 weeks Daily n-3 PUFA intake: 8g
7	Crossover study; 2 2-week periods of either mackerel or herring diet; 3-month interval between dietary periods Subjects: 15 10♂, 5♀, 28-44 years Fish: mackerel Dose: 280 g/day Feeding period: 2 weeks Daily n-3 PUFA intake: 2.2g EPA
12	all subjects consumed fish diet Subjects: 8 with mild essential hypertension Fish: mackerel Dose: 280 g/day Feeding Period: 2 weeks Daily n-3 PUFA intake: 2.2g EPA, 2.8g DHA
2	All subjects consumed fish diet Subjects: 7♂ Fish: mackerel Dose: 500-800 g/day Feeding period: 1 week Daily n-3 PUFA intake: 7-11 g EPA
8	All subjects consumed fish diet Subjects: 3 Fish: mackerel Dose: 750 g/day Feeding period: 3 days Daily n-3 PUFA intake: 10-15 g EPA

10	All subjects consumed control then fish diet
	Subjects: 9 hyperlipidemic
	             5 type IIb
	             4 type V
	Fish: salmon oil and fresh
	Feeding period: 9-28 days

Table 3.1 (continued)

Reference	Experimental design
	I. FATTY FISH

11	All subjects consumed control, then fish diet followed in some subjects by a vegetable oil containing diet
	Subjects: 20 8♂, 12♀ hypertriglyceridemic
	           10 type IIb
	           10 type V
	Fish: salmon, oil and fresh/ or fresh and MaxEPA
	Feeding period: 4 weeks/ 4 weeks
	Daily n-3 PUFA intake: 20 g/ or 30 g

9	All subjects consumed fish diet Subjects: 3 type IIb hyperlipoproteinemics Fish: salmon, fresh, and MaxEPA Feeding period: 4 weeks
4	2 4-week periods of either control or fish diet, 3-week interval between randomly allocated dietary periods Subjects: 11 6♂, 5♀, mean age 39 yrs. Fish: salmon, fresh, and MaxEPA Dose: 1 lb flesh/day, 60-90 ml oil/day Feeding period: 4 weeks Daily n-3 PUFA intake: 10 g
5	3 4-week periods of saturated fat control diet, vegetable oil control diet, or fish diet; 3-week interval between randomly allocated dietary periods Subjects: 12 6♂, 6♀, mean age 40 years Fish: salmon, fresh, and MaxEPA Dose: 1 lb flesh/day, 3-6 Tb. oil/day Feeding period: 4 weeks Daily n-3 PUFA intake: 20-29 g
7	Crossover study; 2 2-week periods of either mackerel or herring diet; 3-month interval between dietary periods Subjects: 15, 10♂, 5♀, 28-44 years Fish: herring Dose: 280 g/day Feeding period: 2 weeks Daily n-3 PUFA intake: 1 g EPA
3	All subjects consumed fish diet Subjects: 10♂, 28-35 years Fish: mackerel and salmon Feeding period: 11 weeks Daily n-3 PUFA intake: 2-3 g EPA

Table 3.1 (continued)

Reference	Experimental design
	I. FATTY FISH
47	All subjects consumed fish diet Subjects: 12 ♂, mean age 26 years Fish: mackerel, salmon, herring Dose: 150 g/day Feeding period: 6 weeks Daily n-3 PUFA intake: 2-3 g EPA
6	Crossover study; subjects randomly divided into two groups; 2 3-month periods of either usual diet or fish diet Subjects: 118 ♂ Fish: fatty fish Dose: 200-600 g/week Feeding period: 3 months
	II. COD LIVER OIL
13	All subjects consumed supplement Subjects: 20 diabetics, mean age: 24.0 ± 1.8 years Dose: 6.8 g/day Feeding period: 2 weeks
14, 15, 16	All subjects consumed supplement Subjects: 6, 35-50 years/ 12 ♂, 19-31 years Dose: 15 ml/ or 20 ml Feeding period: 2 weeks/ or 6 weeks Daily n-3 PUFA intake: 3g/ or 1.8 g EPA, 2.2 g DHA
17	Crossover study; 2 6-week periods of either cod liver oil or corn oil supplementation; 3-week interval between dietary periods Subjects: 10 ♂, 20-30 years Dose: 25 ml/day Feeding period: 6 weeks
18	All subjects consumed supplement Subjects: 7, mean age 32 years Dose: 30 ml/day Feeding period: 2 weeks Daily n-3 PUFA intake: 1.8 g EPA, 1.8 g DHA
19	Crossover study; 2 6-week periods of either cod liver oil or no supplementation Subjects: 17 11 ♂, 6 ♀, hypercholesterolemics, mean age: 36.6 years Dose: 30 ml/day Feeding period: 6 weeks Daily n-3 PUFA intake: 1.5 g EPA

Table 3.1 (continued)

Reference	Experimental design
	II. COD LIVER OIL
8, 20, 21	All subjects consumed supplement Subjects: 8♂, 22-42 years Dose: 40 ml/day Feeding period: 25 days Daily n-3 PUFA intake: 4 g EPA, 6 g DHA
53	All subjects consumed varying doses of supplement during 4-week periods Subjects: 6♂, 26-36 years Dose: 10, 20, 40, 20, 20 ml/day/4 weeks Feeding period: 20 weeks
	III. COD LIVER OIL CONCENTRATE
22,23	Treatment groups (n=10) consumed 1,2,4, and 8 g/day n-3 PUFA; control group (n=10) consumed olive oil Subjects: 52,♂and♀, 18-60 years Dose: 1.58 g, 2.62 g, 4.72 g, 9.45 g/day/treatment group Feeding period: 4 weeks Daily n-3 PUFA intake: 1.37g, 2.27g, 4.09g, 8.19g/group
24	All subjects consumed concentrate Subjects: 20♂, mean age 32 years Dose: 10 ml/day Feeding period: 3 weeks Daily n-3 PUFA intake: 7 g EPA
	IV. SARDINE OIL CONCENTRATE
25	All subjects consumed concentrate Subjects: 7 Dose: 8.1 g/day Feeding period: 4 weeks Daily n-3 PUFA intake: 1.4 g EPA
26	All subjects consumed concentrate Subjects: 8♂, 29-42 years Dose: 8.1 g/day Feeding period: 4 weeks Daily n-3 PUFA intake: 1.4 g EPA, 0.6 g DHA
27	All subjects consumed concentrate Subjects: 12 3♂, 9♀, 31-70 years, hyperlipidemic hemodialysis patients Dose: 9.6 g/day Feeding period: 13 weeks Daily n-3 PUFA intake: 1.6 g EPA, 1.0 g DHA

Table 3.1 (continued)

Reference	Experimental design
	V. MAX EPA
37	All subjects consumed safflower oil supplement, then MaxEPA Subjects: 9 healthy; 2 hyperlipidemic Feeding period: 2-3.5 weeks
47	All subjects consumed MaxEPA Subjects: 8 lactating ♀ Dose: 5 g(n=6)/ 10 g(n=5) Feeding period: 4 weeks(n=6)/ 2 weeks(n=5) Daily n-3 PUFA intake: 1.4 g(n=6)/ 2.8 g(n=5)
33	Randomized doubleblind crossover study, 2 2-week periods of either MaxEPA or olive/maize oil supplementation Subjects: 10 6♂,4♀, 22-35 years Dose: 10 g/day Feeding period: 2 weeks Daily n-3 PUFA intake: 1.7 g EPA, 1.2 g DHA
39	Treatment group consumed MaxEPA; control groups (n=12) received no treatment Subjects: 12♂ Dose: 10 g/day Feeding period: 14 weeks
36	Doubleblind randomized study; subjects assigned to receive MaxEPA or corn/olive oil mix Subjects: 19 15♂,4♀, 56-75 years, with peripheral vascular disease Dose: 10 g/day Feeding period: 7 weeks Daily n-3 PUFA intake: 1.8 g EPA
34	Doubleblind crossover study; 2 4-week periods of either MaxEPA or corn/olive oil mix, 4-week interval between dietary periods Subjects: 20♂, 25-40 years Dose: 10 g/day Feeding period: 4 weeks Daily n-3 PUFA intake: 4 g

Table 3.1 (continued)

Reference	Experimental design
	V. MAX EPA
49	Doubleblind crossover study; 2 2-month periods of either MaxEPA or corn/olive oil mix, 1-month interval between dietary periods Subjects: 16♂ hypertriglyceridemic Dose: 10 g/day Feeding period: 2 months Daily n-3 PUFA intake: 1.8 g EPA
31	All subjects consumed MaxEPA; 2 study groups Subjects: 8, 11 Dose: 10 ml/ or 20 ml/day Feeding period: 12 months Daily n-3 PUFA intake: 1.8g/ or 3.6 g EPA
39	Randomized doubleblind study; control group (n=10) of hypertriglyceridemic received corn/olive oil blend Subjects: 11♂ hypertriglyceridemic Dose: 15 g/day Daily n-3 PUFA intake: 2.7 g EPA; 1.9 g DHA
29	All subjects consumed MaxEPA Subjects: 4, 2♂, 2♀ Dose: 20 ml/day Feeding period: 2 weeks Daily n-3 PUFA intake: 2.5 g EPA; 2.5 g DHA
32	All subjects consumed linseed oil for 2 weeks, then after an interval of 6 weeks consumed MaxEPA Subjects: 5, 3♂, 2♀, 23-30 years Dose: 20 ml/day Feeding period: 2 weeks Daily n-3 PUFA intake: 3 g EPA; 2.9 g DHA
28	All subjects consumed MaxEPA Subjects: 5, 4♂, 1♀, 23-51 years Dose: 20 ml/day Feeding period: 5 weeks
30	All Subjects consumed MaxEPA Subjects: 13, 9♂, 4♀, 33-70 years with ischemic heart disease Dose: 20 ml/day Feeding period: 5 weeks Daily n-3 PUFA intake: 3.5 g EPA

Table 3.1 (continued)

Reference	Experimental design
	V. MAX EPA
40	All subjects consume MaxEPA Subjects: 9, 6♂, 3♀, 26-64 years, on chronic hemodialysis Dose: 20 ml/day Feeding period: 8 weeks Daily n-3 PUFA intake: 3.6 g EPA
35	All subjects consumed MaxEPA Subjects: 92, 31-75 years, with myocardial infarction, angina, xanthelasma, peripheral vascular disease, coronary bypass surgery, diabetes; 15, 31-75 years, healthy Dose: 20 ml/day Feeding period: 1 (n=107), 3 (n=91), 6 (n=72), 9 (n=51), 12 (n=42), and 24 months (n=16) Daily n-3 PUFA intake: 3.6 g EPA
32b	All subjects consumed 3 dosages of MaxEPA in random order for periods of 3 weeks; 6-week interval between experimental periods Subjects: 5♂, 18-31 years Dose: 5, 10, or 20 g/day/feeding period Feeding period: 3 weeks Daily n-3 PUFA intake: 0.8g EPA, 0.8 g DHA (5g); 1.7g EPA, 1.6 g DHA (10g); 3.3g EPA, 3.2 g DHA (20g)
42	All subjects fed experimental diets in varying sequence: control diet, high carbohydrate control diet, and high carbohydrate MaxEPA diet Subjects: 7, 22-54 years, mildly hypertriglyceridemic Dose: 50 g/day Feeding period: 1 week Daily n-3 PUFA intake: 8.5 g EPA, 5.5 g DHA
38	All subjects consumed control and fish oil diet for 4 weeks; 1-month interval between dietary periods Subjects: 7, 6♂, 1♀, mean age: 33 years Dose: 92 g/day, or salmon oil, 120 g/day Feeding period: 4 weeks Daily n-3 PUFA intake: 24 ± 8 g
	VI. PURIFIED EPA
45,46	All subjects consumed EPA Subjects: 5 diabetics Dose: 50 mg EPA arginine salt/day Feeding period: 2 months Daily n-3 PUFA intake: 50 mg EPA

Table 3.1 (continued)

Reference	Experimental design
	VI. PURIFIED EPA
45,48	Treatment group consumed EPA; control group (n=8) consumed placebo Subjects: 11 elderly ♂, mean age: 72 years Dose: 150 mg EPA arginine salt/day Feeding period: 4 weeks Daily n-3 PUFA intake: 150 mg EPA
44	All subjects consumed placebo for 2 weeks, then EPA for 2 weeks Subjects: 12 8♂, 4♀; mean age: 60.7 years Dose: 2 g EPA ethylester [a] capsules/day Feeding period: 4 weeks Daily n-3 PUFA intake: 1.3 g EPA
43	All subjects consumed EPA Subjects: 8♂, 29-42 years Dose: 4.8 g EPA ethylester [b] capsules/day Feeding period: 4 weeks Daily n-3 PUFA intake: 3.6 g EPA

[a] Extracted from fish oil, purity 67%.
[b] Manufactured from sardine oil, purity 75%.

TABLE 3.2 Human Feeding Trials: Summary of Effects of Fish, Fish Oil, or Eicosapentaenoic Acid Consumption on Plasmatic Characteristics

Ref.	Measurements compared to:	Plasmatic characteristics
		I. FATTY FISH
1	Control diet 3 weeks mackerel	TG (mg/dl): 38% TG FA: LA: AA:NSD EPA:↑ DHA:↑ Total Chol.(mg/dl): ↓ 7% Chol.Ester FA(%): LA: ↓ AA:NSD EPA: ↓ DHA:NSD HDL Chol.(mg/dl): ↑7% VLDL Chol.(mg/dl): ↓

7	Before mackerel
	2 weeks mackerel/
	3 months after mackerel
		TG (mmol/L): ↓47% / NSD
		TG FA(vol %): LA:NSD / -; AA:↑50% / -;
		  EPA:↑425% / -; DHA:↑1016% / -
		Total Chol.(mmol/L): ↓ 7% / NSD
		Chol.Ester FA(vol %): LA: ↓20% / -;
		  AA:NSD /-; EPA: 626% /-; DHA:↑171% /-;
		HDL-, LDL Chol.:NSD

12	Before mackerel 2 weeks mackerel	TG FA: LA:NSD; AA:NSD; EPA:↑800%; DHA:↑711% Chol.Ester FA: LA:↓24%; AA:NSD; EPA:↑636%; DHA:↑137%
2	Before mackerel 3 days mackerel/ 6 days mackerel	TG: NSD / NSD Total Chol.: NSD / NSD
8	Before mackerel 1 day mackerel/ 3 days mackerel	Lipid FA: LA:↓63% / ↓74%; AA:NSD / NSD; EPA:↑580% / ↑1009%; DHA:NSD / NSD
10	Control diet Salmon diet	TG(mg/dl): Type IIb:↓65%; Type V:↓50% Total Chol.(mg/dl):Type IIb:↓24%; Type V:↓13% LDL Chol.(mg/dl):Type IIb:↓15%; Type V:↑104%
11	Control diet Fish oil diets	TG(mg/dl): Type IIb:↓64%; Type V:↓79% Total Chol.(mg/dl): Type IIb:↓27%; Type V:↓48% HDL Chol.(mg/dl): Type IIb:↓17%; Type V:NSD LDL Chol.(mg/dl): Type IIb:↓12%; Type V:↑48% VLDL Chol.(mg/dl): Type IIb:↓73%; Type V:↓71%
9	Control diet Salmon diet	TG(mg/dl): 67%↓ Total Chol.(mg/dl):↓20%

Table 3.2 (continued)

Ref.	Measurements compared to	Plasmatic characteristics
		I. FATTY FISH
5	Control diet Salmon diet	Lipid FA: LA:↓50%; AA:↓23%; EPA:↑; DHA:↑ TG(mg/dl): ↓ 38% Total Chol.(mg/dl):↓14% HDL Chol.:NSD; LDL Chol.(mg/dl): ↓ 16% VLDL Chol.(mg/dl): ↓ 38%

7	Before herring
	2 weeks herring/
	3 months after
	herring
		TG:NSD / NSD
		TG FA(vol %): LA:NSD / -; AA: 40% / -;
		   EPA:NSD / -; DHA:↑485% / -;
		Total Chol.: NSD / NSD
		Chol. Ester FA(vol %):LA: 10% /-; AA:NSD /-;
		   EPA: 332% /-; DHA: ↑ 100% / -;
		HDL-, LDL Chol.: NSD / NSD

6	Usual diet Fatty fish diet	TG(mmol/l):↓6% Total Chol.:NSD HDL-,LDL Chol.: NSD
		II. COD LIVER OIL
13	Before supplem. 2 weeks supplem.	Lipid FA(%): LA:NSD; AA:NSD; EPA:↑65%; TG:NSD Total Chol.:NSD Chol.Ester FA:LA:NSD; AA:NSD; EPA:↑53%: DHA:- HDL Chol.:NSD

14	Before supplem.
	1 week supplem./
	2 weeks supplem./
	1 wk after suppl./
	2 wks after suppl./
		Lipid FA (phospholipid; mole %):
		   LA: ↓ 15% / ↓ 21% / NSD / NSD
		   AA: NSD / NSD / ↓ 20% / NSD
		   EPA: ↑ 288% / ↑ 375% / ↑ 50% / NSD
		      DHA: ↑ 95% / ↑ 109% / ↑ 90% /↓
		TG FA(mole %):
		   LA: NSD / NSD / NSD / NSD
		   EPA: ↑ / ↑ / NSD / NSD
		Chol.Ester FA:
		   LA: NSD / ↓ 7% / NSD NSD
		   EPA: ↑ 300% / ↑ 343% / ↑ 71% / NSD

15, 16	Before supplem. At end of suppl./ 5 wks after end of supplem.	TG(mmol/L): ↓ 22% / NSD Total Chol.: NSD / NSD HDL Chol.(mmol/L): ↑ 9% / NSD Antithrombin III(%): ↓ 16% / ↓ 20%
17	Before supplem. 6 wks CLO suppl./ 16 wks after end of supplem.	Lipid FA(mole%): LA: NSD / -; AA: NSD / -; EPA: ↑ 233% / -; DHA: 114% / - TG: NSD / - Total Chol.: NSD / -

Table 3.2 (continued)

Ref.	Measurements compared to	Plasmatic characteristics
		II. COD LIVER OIL
19	Before supplem. 6 wks CLO suppl./ 6 wks after CLO supplem.	Lipid FA(mole %): LA: ↓ / -; EPA: ↑ 100% / - TG: NSD Total Chol.: NSD HDL Chol.: NSD
21	Before supplem. 25 days suppl.	Lipid FA(rel %): LA: ↓ 39%: AA: NSD EPA: 829%; DHA: ↑ 118% TG: NSD Total Chol.:NSD HDL, LDL: NSD
53	Before supplem. 4 wks (10ml/day)/ 4 wks (20ml/day)/ 4 wks (40ml/day)/ 4 wks (20ml/day)/ 4 wks (20ml/day)/ 8 wks after supp./ 20 wks after supplem.	Lipid FA (rel %): LA: NSD/-/NSD/-/NSD/-/NSD AA: NSD/-/NSD/-/NSD/-/NSD EPA: NSD/-/↑/-/NSD/-/NSD DHA: NSD/-/↑/-/NSD/-/NSD Lipid FA (phospholipid; rel %): LA: ↓/↓/↓/↓/↓/NSD/NSD AA: NSD/NSD/↓/↓/↓/NSD/NSD EPA: ↑/↑/↑/↑/↑/NSD/NSD DHA: ↑/↑/↑/↑/↑/↑/NSD TG(mg%): NSD/↓/↓/↓/↓/NSD/NSD Total Chol.(mg%):NSD/NSD/NSD/NSD/NSD/NSD/NSD HDL Chol.(mg %):NSD/NSD/NSD/NSD/NSD/NSD/NSD
		III. COD LIVER OIL CONCENTRATES
22, 23	Before supplem. End of supplem./ 2 wks after end of supplem.	Lipid FA(%): LA: ↓ / NSD; AA: NSD / NSD; EPA: ↑ / NSD; DHA: ↑ / NSD TG(mg/100ml; with 8g n-3 PUFA: ↓ 39% Total Chol.: NSD / - HDL Chol.: NSD / -; HDL:↓(w/8g n-3 PUFA)
24	Before supplem. 3 weeks supplem.	Antithrombin III(g/L): 9%
		IV. SARDINE OIL CONCENTRATES
26	Before supplem. 2 wks supplem./ 4 wks supplem	Lipid FA(ug/ml): AA: NSD/NSD; EPA: ↑ 158% / ↑ 144%; DHA: NSD / NSD TG: NSD / NSD Total Chol.: NSD / NSD HDL Chol.: NSD / NSD
27	Before supplem. 4 wks supplem./ 13 wks supplem.	Lipid FA (%): LA: ↓ 6% / ↓ 13%; AA: NSD / NSD; EPA: ↑ 266% / ↑ 342%; DHA: ↑ 42% / ↑ 53% TG: ↓ / ↓ Total Chol.: ↓ / ↓ HDL Chol. NSD / NSD

Effects of Omega-3 Polyunsaturated Fatty Acid

Table 3.2 (continued)

Ref.	Measurements compared to	Plasmatic characteristics
		V. MAX EPA
37	Before supplem. Last 5 days MaxEPA	Lipid FA(%): LA: ↓ ; AA: NSD; EPA: ↑ ; DHA: ↑ TG: ↓ HDL Chol.: ↓ LDL Chol.: ↑ VLDL Chol.: ↓ HDL: ↓ ; VLDL: ↓

41	Before supplem.
	2 wks 5g/day /
	1 wk 10g/day /
	4 wks 5/g day /
	2 wks 10g/day
		Lipid FA (phospholipid):
		  LA: - / - / NSD / NSD
		  AA: - / - / NSD / NSD
		  EPA: - / - / ↑ / ↑
		  DHA: - / - / ↑ 400% / ↑ 730%

32	Before supplem. 2 wks Max EPA	TG (mmol/L): ↓ Total Chol.: NSD HDL Chol.(mmol/L): ↑
39	Before supplem. 3 wks Max EPA/ 5 wks Max EPA/ 8 wks Max EPA/ 14 wks Max EPA	TG (mmol/L): ↓ / ↓ / ↓ / ↓ HDL Chol.(mmol/L): ↑ / ↑ / ↑ / ↑
36	Before supplem. 7 wks Max EPA	TG(mmol/L): ↓ 27% Total Chol.: NSD HDL Chol.: NSD
34	Before supplem. 4 wks Max EPA	TG (mmol/L): ↓ 26% Total Chol.: NSD HDL Chol.: NSD HDL, LDL: NSD VLDL (g/L): 30% Antithrombin III: ↑ 5.5%
49	Before supplem. 2 months Max EPA	TG (mmol/L): ↓ 48% Total Chol.(mmol/L): ↓ 8% HDL Chol. (mmol/L): ↑ 17%
31	Before supplem. 1 month supplem./ 12 months supplem.	TG (mmol/L): ↓ 36% / ↓ 38% Total Chol.: NSD / NSD HDL Chol.(mmol/L): ↑ 9% / NSD
39	Before supplem. After MaxEPA	TG (mmol/L): ↓ 43% Total Chol.: NSD HDL Chol.: NSD

Table 3.2 (continued)

Ref.	Measurements compared to	Plasmatic characteristics
		V. MAX EPA
29	Before supplem. 2 wks Max EPA	Lipid FA(Plasma Choline Phosphoglycerides wt%) LA: ↓ ; EPA: ↑346%; DHA: ↑15%
32	Before supplem. 2 wks Max EPA	TG (mmol/L): ↓43% Total Chol.: NSD HDL Chol.: NSD
28	Before supplem. 5 wks Max EPA	TG (mmol/l): ↓35% Total Chol.: NSD HDL Chol.(mmol/L): ↑22%
30	Before supplem. 5 weeks Max EPA	TG: NSD Total Chol.: NSD LDL(mmol/L): ↑18%
40	Before supplem. 8 weeks Max EPA	TG(mmol/L): ↓51% Total Chol.: NSD HDL Chol.(mmol/L): 25%
35	Before supplem. 1 month Max EPA/ 3 mons. Max EPA/ 6 mons. Max EPA/ 12 mons. Max EPA/ 24 mons. Max EPA	TG: 37% / ↓36% / ↓41% / ↓38% / ↓37% / ↓41% Total Chol.: NSD/ ↓2%/↓4%/NSD/NSD/↓5% HDL Chol.: ↑10%/ ↑7%/ ↑10%/NSD/NSD/ ↑14%
32b	Before supplem. 3 wks. (5g/day)/ 3 wks. (10g/day)/ 3 wks. (20g/day)/	TG(mmol/L): ↓14%/↓23%/↓32% Total Chol.(mmol/L): NSD/NSD/↓9% HDL Chol.(mmol/L): NSD/NSD/ ↑31%
42	Before supplem. After 1 wk suppl.	TG: ↓ Total Chol. ↓
38	Before supplem. last 2 weeks	Phospholipid FA(% total FA): LA: ↓80%; AA: ↓20%; DHA: 623% TG(mg/dl): ↓43% Total Chol.(mg/dl): ↓23% Chol. Ester FA(% total FA): LA: 68%; AA: NSD; DHA: ↑2600% HDL Chol.: NSD LDL Chol. (mg/dl): ↓20% VLDL(mg/dl): ↓52%
		VI. PURIFIED EPA
46, 45	Before supplem. 2 mons. supplem.	Lipid FA: LA,AA,EPA,DHA: NSD

Table 3.2 (continued)

Ref.	Measurements compared to	Plasmatic characteristics
		VI. PURIFIED EPA
48	Before supplem. 4 weeks supplem.	Lipid FA: LA,AA,EPA,DHA: NSD
44	Before supplem. 2 weeks supplem./ 4 weeks supplem.	Lipid FA: AA: NSD/NSD; EPA: ↑/↑; TG(mg/dl): NSD/ NSD Total Chol.(mg/dl): NSD /↓ HDL-, LDL Chol.: NSD/NSD
43	Before supplem. 2 weeks supplem./ 4 weeks supplem.	Lipid FA: AA:NSD/NSD; EPA: ↑/↑; TG(mg/dl): - / ↓ 26% Total Chol.: - / NSD HDL Chol.: - / NSD

NSD: no significant difference ($p \leq 0.05$) between comparison times
↑: increase
↓: decrease

TABLE 3.3 Human Feeding Trials: Summary of Effects of Fish, Fish Oil, or Eicosapentaenoic Acid Consumption on Platelet Characteristics

Ref.	Measurements compared to	Platelet characteristics
		I. FATTY FISH
2	Before mackerel 3 days mackerel/ 6 days mackerel	Phospholipid FA(ug/mg): LA: -/↓41%; AA:-/↓28% EPA: -/↑244%; DHA: -/↑195% Aggregation (PRP) Induced by: AA(1.8 nmol/L): NSD/NSD Adrenalin(500umol/L): NSD/NSD; Collagen(1ug/ml):↓39%/↓49%; Collagen(10ug/ml): NSD/NSD TXB2 Production (PRP) Induced by: AA(1.8nmol/L): NSD/NSD Adrenalin(500uM/L): NSD/NSD Collagen(1ug/ml): NSD/↓33% Collagen(10ug/ml): NSD/NSD
4	Control diet Salmon diet	Phospholipid FA(%FA): LA:↓47%; AA:↓27%; EPA: ↑6000%; DHA: ↑236% Count: ↓34% Aggregation (PRP) Induced by: AA: NSD; Adrenalin:NSD ADP(5uM):NSD; ADP(2uM): ↓24%; ADP(1.4uM): ↓33% Collagen(0.2mg/ml):NSD Thrombin: NSD Bleeding time: ↑48% MDA Production(PRP) Induced by N-ethyl-maleimide: ↑35%
3	Before fish diet 3 wks fish diet/ 6 wks fish diet/ 11 wks fish diet/ 6 wks fish diet/ 11 wks fish diet	Phosphatidylcholine(%FA): LA: ↓14%/ ↓15%/ ↓17%/ NSD/ NSD AA: NSD/ NSD/ NSD/ NSD/ ↑9% EPA: ↑200%/↑250%/↑300%/NSD/NSD DHA: ↑47%/ ↑47%/ ↑59%/NSD/NSD Count: NSD/NSD/NSD/NSD/NSD Aggregation (PRP) Induced by: Adrenalin: NSD/ NSD/ NSD/ -/- Smallest conc. ADP: NSD/↑/↑/↑/↑ Collagen(1ug/ml):↓/-/-/-/- Collagen(3ug/ml):↓/-/-/-/- Collagen(4ug/ml):↓/-/-/-/- Bleeding time: ↑/↑42%/↑33%/NSD/NSD

Table 3.3 (continued)

Ref.	Measurements compared to	Platelet characteristics
		I. FATTY FISH

47	Before fish diet
	1 wk fish diet/
	3 wks fish diet/
	6 wks fish diet/
	
	3 wks after fish/
	8 wks after fish/
	14 wks after fish
		Phosphatidylcholine(%):
		    LA: ↓ 13%/ ↓ 7%/ ↓11%/ NSD/ NSD/ -
		    AA: ↓ 7%/ ↓ 11%/ ↓ 15%/ ↓ 6%/ NSD/ -
		    EPA: ↑ 216%/ ↑ 216%/ ↑ 233%/ ↑ 50%/ ↑ 50%/
		    DHA: ↑ 40%/ ↑ 53%/ ↑ 66%/ ↑ 27%/ ↑ 20%/-
		Count: NSD/ NSD/ NSD/ NSD/ NSD/ NSD
		Aggregation (PRP) Induced by:
		    Adrenalin: NSD/ ↓ / ↓ / ↓ / ↓ / ↓
		    Collagen(10ug/ml): ↓ / ↓ / ↓ / ↓ / ↓ / ↓
		Bleeding Time: NSD/ NSD/ ↑ 37%/ ↑ / NSD/ NSD

		II. COD LIVER OIL
13	Before supplem. 2 wks supplem.	Aggregation (PRP) Induced by: ADP(5uM): subjects with initial abnormal agregation: reached normal levels; subjects with normal aggregation: remained unchanged

15, 16	Before supplem.
	Weekly during/
	At end of suppl./
	5 wks after end
		Lipid FA(g/100g methyl esters):
		    LA: -/NSD/NSD; AA: -/↓15%/NSD;
		    EPA: -/↑433%/↑83%; DHA:-/↑76%/↑23%
		Count: -/NSD/NSD
		Bleeding time: ↑ / ↑40%/ NSD

17	Before supplem.
	6 wks supplem./
	16 wks after end
	of supplem.
		Phosphatidylcholine (mole%):
		    LA: NSD/-; AA: ↓ 18%/-
		    EPA: ↑ 166%/-; DHA: ↑ 60%/-
		Count: NSD/ -
		Aggregation (PRP) Induced By:
		    AA(313uM): NSD/-; AA(625uM):NSD/-;
		    AA(937uM): NSD/-;
		    Collagen(0.9ug/ml):↓/NSD;
		    Collagen(1.3ug/ml): ↓/NSD;
		    Collagen(2.5ug/ml): ↓ /NSD
		TXB2 Production (PRP) Induced By:
		    Adrenalin: NSD
		    Collagen(0.9ug/ml): ↓ / ↓
		    Collagen(1.3ug/ml): ↓ /NSD
		    Collagen(2.5ug/ml): ↓ /NSD
		Bleeding Time: NSD/-
		MDA Production (PRP) Induced By:
		    AA(313uM):↑/-; AA(625uM): ↑/-;
		    AA(937uM):↑/-; Collagen:NSD/-

Table 3.3 (continued)

Ref.	Measurements compared to	Platelet characteristics
		II. COD LIVER OIL
18	Before supplem. 2 wks supplem./ 2 wks after end of supplem.	Phosphatidylcholine (wt.%): LA: ↓23%/NSD; AA: ↓ 32%/NSD; EPA: ↑867%/NSD; DHA: ↑ 157%/↑86%; Aggregation (PRP) Induced By: Thrombin(0.8U/ml): ↓ /- Bleeding Time: ↑ 81%/NSD
19	Before supplem. 6 wks supplem./ 16 wks after supp.	Phospholipids (%): AA: ↓ /-; EPA: ↑/- Count: NSD/- Aggregation (PRP) Induced By: ADP: NSD/-; Collagen(0.2ug/ml): ↓ /↓; Collagen(>0.2ug/ml): NSD/- Thrombin (0.1, 0.15U/ml): NSD/- TXB2 Production (PRP) Induced By: Adrenalin (0.2ug/ml): NSD/- Adrenalin (>0.2 ug/ml): NSD/- Thrombin(0.05U/ml): NSD/-; Thrombin(0.10U/ml): NSD/-; Thrombin(0.15U/ml): NSD/-; Thrombin(0.25U/ml): ↓ /NSD; Thrombin(0.50U/ml): NSD/- MDA Production (PRP) Induced By: Collagen: NSD/- Thrombin: NSD/-
8, 20, 21	Before supplem. 25 days supplem.	Lipid FA (% total FA): LA:NSD; AA: ↓ 24%; EPA:↑1467%; DHA:↑113% Count: ↓ 11% Aggregation (PRP) Induced By: AA(0.5-2.5uM):NSD ADP(<1.5uM):↓; ADP(>1.5uM):NSD Collagen(0.25ug/ml):↓65% Collagen(0.75ug/ml):↓24% Collagen(10.0ug/ml):NSD TXB2 Production (PRP) Induced By: AA:NSD Collagen(0.25ug/ml):↓45% Collagen(0.75ug/ml):↓63% Bleeding Time: ↑39%

Table 3.3 (continued)

Ref.	Measurements compared to	Platelet characteristics
		II. COD LIVER OIL
53	Before supplem. 4 wks(10ml/day)/ 4 wks(20ml/day)/ 4 wks(40ml/day)/ 4 wks(20ml/day)/ 4 wks(20ml/day)/ 8 wks after suppl/ 20 wks after suppl.	Phospholipid FA (rel%): LA:NSD/NSD/NSD/NSD/NSD/NSD/NSD AA:NSD/↓/↓/↓/↓/↓/NSD EPA:NSD/↑/↑/↑/↑/NSD/NSD DHA:NSD/↑/↑/↑/↑/↑/NSD Count: NSD/NSD/NSD/NSD/NSD/NSD/NSD Aggregation (PRP) Induced by: Collagen(0.75ug/ml):-/↓/↓/-/↓/-/NSD Collagen(0.25ug/ml):-/↓/↓/-/↓/-/NSD
		III. SARDINE OIL CONCENTRATE
26	Before supplem. 2 wks supplem./ 4 wks supplem.	Lipid FA(ug/10^9): AA:NSD/NSD; EPA: ↑128%/↑167% DHA: NSD/NSD Aggregation (PRP) Induced By: ADP(2umol/L): NSD/NSD Collagen(1ug/ml): ↓37%/↓57%
27	Before supplem. 4 wks supplem./ 13 wks supplem.	Count: -/↓19%
		IV. MAX EPA
33	Before supplem. 2 wks Max EPA	Aggregation (PRP) Induced By: ADP(2uM): NSD Collagen(0.5ug/ml): ↓ Collagen(1.0ug/ml):NSD Collagen(10.0ug/ml):NSD U46619(1.5uM):NSD TXB2 Production (PRP) Induced By: Collagen (0.5ug/ml): ↓ Collagen (1.0ug/ml): ↓ Collagen(10.0ug/ml): ↓
36	Before supplem. 7wks Max EPA	Count: NSD
34	Before supplem. 4 wks Max EPA	Lipid FA(%): LA: ; AA: ; EPA:↑; DHA:↑ Count: NSD Aggregation (PRP) Induced By: ADP: NSD; Collagen:NSD Bleeding Time: 16%
31	Before supplem. 1 mon. supplem./ 12 mon. supplem.	Count: 8%/NSD Bleeding Time (min): 10 ml:NSD/NSD; 20ml: NSD/ ↑100%

Table 3.3 (continued)

Ref.	Measurements compared to	Platelet characteristics
		IV. MAX EPA
29	Before supplem. 2 wks supplem.	Phosphoglycerides (wt%): AA:↓; EPA:↑241%; EHA:↑44%

32	Before supplem.
	2 wks supplem.
		Phosphoglycerides (wt%):
		    LA:NSD; AA: ↓ 22%; EPA: ↑ 355%;
		    DHA:↑54%

28	Before supplem. 5 wks supplem.	Count: 16%

40	Before supplem.
	8 wks supplem.
		Aggregation (PRP) Induced by:
		    ADP(0.8uM):↓; ADP(2.0uM):↓;
		  ADP(5.0uM):↓
		  Collagen(0.5ug/ml):NSD
		  Collagen(1.0ug/ml):↓
		  Collagen(2.0ug/ml):↓

35	Before supplem. 1 mon. supplem./ 3 mon. supplem./ 6 mon. supplem./ 9 mon. supplem./ 12 mon. supplem./ 24 mon. supplem.	Count: ↓ 9%/ ↓ 9%/ ↓ 8%/ NSD/ NSD/ NSD

32b	Before supplem.
	3 wks 5g/day
	3 wks 10g/day
	3 wks 20g/day
		Lipid FA (wt%):
		  LA: NSD/ NSD/ NSD
		  AA: ↓13%/ ↓20%/ ↓20%
		  EPA: ↑155%/ ↑255%/ ↑366%
		  DHA: ↑29%/ ↑38%/ ↑79%
		Aggregation (PRP) Induced by:
		  Collagen (1ug/ml): NSD/NSD/NSD
		  Collagen (2ug/ml): NSD/NSD/NSD
		  Collagen (10ug/ml): NSD/NSD/NSD
		  U46619(3uM): NSD/ NSD/ NSD
		Bleeding Time(min): ↑26%/ ↑11%/ ↑25%

		V. PURIFIED EPA

46,
45
	Before supplem.
	2 mon. supplem.
		Lipid FA: LA:NSD; AA:NSD; EPA:NSD; DHA:NSD
		Aggregation (PRP) Induced by:
		  AA:NSD; Collagen: ↓ 50%
		TXB2 Production (PRP) Induced by:
		  Thrombin: NSD

Table 3.3 (continued)

Ref.	Measurements compared to	Platelet characteristics
		V. PURIFIED EPA
45, 48	Before supplem. 4 wks supplem.	Phosphatidylcholine(wt%): LA:NSD; AA:NSD; EPA:NSD; DHA:NSD Aggregation (PRP) Induced by: Adrenalin(0.25uM): ↓35%; Adrenalin(1.0uM): ↓ 33%; Adrenalin(5uM): ↓ 19% ADP(1uM): ↓43%; ADP(2uM): 40%; ADP(3uM):NSD; ADP(5uM):NSD Collagen(40ug/ml): ↓ 15% TXB2 Production (PRP) Induced by: Thrombin: ↓60%
44	Before supplem. 2 wks supplem./ 4 wks supplem.	Phospholipids: AA:NSD/NSD; EPA ↑/↑ Aggregation (PRP) Induced by: Adrenalin(1uM): ↓/↓ Collagen(2ug/ml): ↓/↓ Bleeding Time: NSD/NSD
43	Before supplem. 2 wks supplem./ 4 wks supplem.	Lipid FA: LA: -/NSD; AA: -/NSD; EPA: -/ 128%; DHA: -/ NSD Aggregation (PRP) Induced by: AA: NSD/ NSD; Adrenalin (1ug/ml): NSD/ ↓49% ADP(2uM): NSD/ ↓56% Collagen(1ug/ml): ↓88%/ ↓91%

Comparisons:

    NSD: no significant difference ($p \leq 0.05$) between comparison times
    ↓ : significant decrease between comparison times
    ↑ : significant increase between comparison times

TABLE 3.4 Human Feeding Trials: Summary of Effects of Fish, Fish Oil, or Eicosapentaenoic Acid Consumption on Blood and Urinary Characteristics

Ref.	Measurements compared to	Blood and urinary characteristics
		I. FATTY FISH
7	Before mackerel 2 weeks mackerel 3 mon. after mackerel	Blood Pressure(mmHg): systolic: ↓12%/ NSD diastolic: ↓9%/ ↓9%
12	Before mackerel 2 weeks mackerel	Blood pressure(mmHg): systolic: ↓8%; diastolic: NSD
8	Before mackerel 1 day mackerel 3 days mackerel	Urinary 2,3 dinor-6-keto PGF_1(ng/24hrs) : ↑59%/ ↑59%; Δ-17-2-3-dinor-6-keto-PGF_1(ng/24hrs): ↑ / ↑
		II. COD LIVER OIL
15, 16	Before supplem. at end of suppl. 5 weeks after end	Blood pressure(mmHg): systolic: ↓10%/ ↓7%; diastolic: ↓15%/ ↓12%
17	Before supplem 6 weeks supplem. 16 weeks after end	Hematocrit: NSD/ - Vessel wall PGI2: NSD/-; 6-keto $PGF_1\alpha$: NSD/-; Δ-17-6-keto $PGF_1\alpha$: NSD/- Urinary PGE + PGF metabolites: NSD/- Δ-17-2-3-dinor-6-keto $PGF_1\alpha$: ↑
19	Before supplem 6 weeks CLO	Hematocrit: NSD
8, 20, 21	Before supplem. 25 days supplem.	Hematocrit: NSD Hemoglobin: NSD Blood pressure(mmHg; upright/supine): systolic: ↓/NSD; diastolic: NSD/NSD Urinary 2,3 dinor-6-keto $PGF_1\alpha$: NSD
		III. COD LIVER OIL CONCENTRATE
22, 23	Before supplem. Weekly during suppl. 2 weeks after end	Hemoglobin: NSD/NSD

Table 3.4 (continued)

Ref.	Measurements compared to	Blood and urinary characteristics
		IV. SARDINE OIL CONCENTRATE
27	Before supplem. 4 weeks supplem. 8 weeks supplem. 12 weeks supplem.	Blood and Plasma viscosity: NSD/NSD/NSD Blood pressure: systolic: NSD/NSD/NSD diastolic: NSD/ NSD/ ↓ 9%
		V. MAX EPA
36	Before supplem. 7 weeks supplem.	Hemoglobin: NSD Blood viscosity: ↓ ; Plasma viscosity: NSD
34	Before supplem. 4 weeks supplem.	Hemoglobin: NSD Blood pressure: systolic: ↓ 4%; diastolic: NSD
40	Before supplem. 8 weeks supplem.	Hemoglobin: NSD
		VI. PURIFIED EPA
43	Before supplem. 2 weeks supplem. 4 weeks supplem.	Hematocrit: NSD/NSD Blood viscosity: -/↓; Plasma viscosity: -/NSD Blood pressure: systolic and diastolic: NSD/NSD

Comparisons:
NSD: no significant difference ($p \leq 0.05$) between comparison times
↓ : significant decrease between comparison times
↑ : significant increase between comparison times

TABLE 3.5 Human Feeding Trials: Summary of the Effects of Fish Oil and Eicosapentaenoic Acid Consumption on Erythrocyte Characteristics

Ref.	Measurements compared to	Erythrocyte characteristics
		I. COD LIVER OIL
14	Before supplem. 1 wk supplem./ 2 wks supplem./ 1 wk after suppl./ 2 wks after suppl.	Phosphatidylcholine(mole%): LA: ↓ 6%/ ↓ 13%/ ↓ 3%/NSD; AA: NSD/ NSD/ NSD/ NSD EPA: ↑ 275%/ ↑ 450%/ ↑ 125%/ ↑ 50% DHA: ↑ 33%/ ↑ 100%/ ↑ 75%/ ↑ 58%
15, 16	Before supplem. weekly during/ at end of suppl./ 5 wks after end	Lipid FA (wt%): LA: -/ ↓ 13%/ NSD AA: -/ ↓ 12%/ ↓ 12% EPA: -/ ↑ 236%/ ↑ 164% DHA: -/ ↑ 43%/ ↑ 47% Count: NSD
8, 20, 21	Before supplem. 25 days supplem.	Lipid FA (rel.%): LA:↓; AA: NSD; EPA:↑; DHA:↑ Count: NSD
53	Before supplem. 4 wks 10ml/day 4 wks 20ml/day 4 wks 40ml/day 4 wks 20ml/day 4 wks 20ml/day 8 wks after suppl./ 20 wks after suppl.	Phospholipids (rel.%): LA: ↓/↓/↓/↓/↓/NSD/NSD AA: ↓/↓/↓/↓/↓/NSD/NSD EPA: NSD/↑/↑/↑/↑/NSD/NSD DHA: ↑/↑/↑/↑/↑/↑/↑
		II. PURIFIED EPA
43	Before supplem. 2 wks supplem./ 4 wks supplem.	Phospholipids (mole%): LA:-/NSD; AA:-/NSD; EPA:-/ ↑107%; DHA: -/NSD

Comparisons:

    NSD: no significant difference ($p \leq 0.05$) between comparison times
    ↓ : significant decrease between comparison times
    ↑ : significant increase between comparison times

EPA ingested, which was reported more often than that of DHA, ranged from 1.5 [19] to 4 g per day [8,20,21]. The fatty acid composition of the cod liver oil fed to subjects in four of the studies is presented in Table 3.6. EPA constituted 8 to 10% of the fatty acids present; the corresponding values for DHA were 11 to 13.8%. Fatty acids such as 16:0, 16:1, 18:1, and 20:1 were present in greater amounts, while the concentration of 22:5n-3 was very low.

Concentrated n-3 PUFAs prepared from cod liver [22-24] and sardine oils [25-27] were used in other trials. The fatty acid composition and mode of preparation of the concentrates fed to the subjects is presented in Table 3.7. DHA and EPA were the most abundant fatty acids in the cod liver oil concentrate. EPA was the fatty acid present in highest amount in the sardine oil concentrate, while 16:0, 16:1, and 18:1 were more abundant than DHA. 22:5n-3 was present in very small amounts in both oil concentrates.

An n-3 PUFA concentrate manufactured by British Cod Liver Oils (Hull, England) and marketed under the tradename MaxEPA containing 0.18 g of EPA and 0.12 g of DHA/g oil [50] was also studied. The daily amounts consumed varied between 5 g [32b] and 92 g [38] for periods ranging from one week [42] to twelve months [31].

Eicosapentaenoic acid, in the form of a purified arginine salt [45,46,48] or its ethylester [43,44], was also used as a dietary supplement. The amount consumed daily ranged from 50 mg [45,46] to 3.6 g [43] in trials lasting four weeks [43,44,45] and two months [45,46].

Feeding Trials: Subjects

Most of the feeding trials were conducted with healthy adults of either or both sexes as subjects. Patients with diabetes [45,46], serum lipid abnormalities [9,10,11,19,27,37,39,49], peripheral vascular disease [35,36,55], ischemic heart disease [30], and angina [35] have also been included. Individuals recovering from myocardial infarctions [35] or requiring chronic hemodialysis [27,40] also participated.

The effects of fish or fish oil consumption on selected plasma, platelet, and erythrocyte characteristics, prostacyclin production by

TABLE 3.6 Fatty Acid Composition of Cod Liver Oil Fed to Subjects

Fatty Acid	Weight %		
	British Cod Liver Oil[a]	Collett Marwell Hauge[a]	Lovitran[a]
14:0	3.7	4	4.7
14:1	-	-	0.4
16:0	10.7	9	10.8
16:1	8.7	12	11.3
18:0	2.7	2	2.1
18:1	22.4	24	25.7
18:2n-6	0.7	2	2.3
18:3	0.5	-	-
18:4n-3	-	3	2.1
20:0	1.3	-	-
20:1n-9	-	13	-
20:1n-11	11.5	-	13
20:4n-6	-	-	-
20:5n-3	10.3	8	8.8
22:1n-9	-	6	-
22:1n-11	7.0	-	6.8
22:5n-3	1.7	2	-
22:6n-3	12.5	11	11.3
Others	-	4	0.7
Reference	16	17,19	14

[a] Brand name

TABLE 3.7 Fatty Acid Composition of Cod Liver and Sardine Oil Concentrate Fed to Subjects

Fatty acid	Weight %		
	Cod liver oil concentrate	Sardine oil concentrate	
14:0	1.0	-	6.1
16:0	-	10.8	13.4
16:1	1.3	11.9	10.4
16:2	2.0	-	-
18:0	-	-	1.8
18:1	2.0	14.9	13.8
18:2	1.6	4.8	3.4
18:3n-3	1.0	0.6	-
18:4n-3	10.0	-	3.6
20:1	-	3.1	6.0
20:3n-3	0.7	-	-
20:3n-6	-	0.2	-
20:4n-6	1.6	0.8	1.0
20:5n-3	27.6	21.3	16.8
22:1	-	-	5.0
22:4	2.8	-	-
22:5n-3	1.2	-	3.2
22:6n-3	44.6	8.4	10.6
24:1	-	-	2.0
Reference	(22,23)[a]	(26)[b]	(27)[c]

[a] Preparation included: saponification of oil, separation of PUFAs from saturated and monounsaturated fatty acids with saturated solution of urea in methanol, extraction of PUFAs with hexane, hexane removal and packaging into gelatin capsules

[b] Prepared according to the Boucher method with minor modification, presented as gelatin-coated capsules

[c] Obtained by winterizing crude oil and encapsulating in gelatin

vessel walls, and the levels of urinary prostaglandin metabolites have been monitored. These are reviewed here, followed by a discussion of the research methodologies used and some questions that remain to be answered by further long-term studies.

EFFECTS OF FATTY FISH, MARINE OIL, OR n-3 PUFA CONSUMPTION ON PLASMATIC CHARACTERISTICS

Plasma Lipid Fatty Acids

Changes in the fatty acid composition of plasma lipids occurred in all feeding trials (Table 3.2). Decreases in the level of LA and increases in EPA were observed when participants consumed diets rich in fatty fish [8], or when the subjects' usual diets were supplemented with cod liver oil [8,14,20,21,53], cod liver [22,23] or sardine oil concentrates [27], or MaxEPA [29,37,38]. Increases in the plasma level of DHA were observed in feeding trials using cod liver oil [8,14,17,20,53], cod liver [22,23] and sardine oil concentrates [27], and MaxEPA [29,37] but not purified EPA [43]. In those trials where the values for the changes in both n-3 PUFA levels were given, the increase in the amount (percent, mole percent, relative percent, or weight percent) of EPA was usually several times greater than that observed for DHA.

In general, the AA content of plasma lipids appeared more resistant to change; for example, AA decreased 20% in a feeding trial using MaxEPA [38] but not where lesser amounts of the supplement were consumed [37]. In a 20-week feeding trial using cod liver oil supplementation, AA decreased only in subjects consuming the highest doses (40 ml/day) [53]. There were no significant changes in AA levels when sardine [26,27] and cod liver oil [22,23] concentrates or purified EPA [44,45] were used. The plasma lipid fatty acid changes occurring with cod liver oil and sardine oil concentrate consumption are shown in Tables 3.8 and 3.9.

Changes in the relative content of fatty acids in plasma lipids became evident very soon after the initiation of the experimental period. For example, a 580% increase in the relative percentage of EPA in plasma phospholipids and a 63% decrease in the LA content were observed after only one day on a mackerel diet [8]. In a more controlled study, increases in the percentages of n-3 PUFAs in plasma lipids were observed after only one day on a salmon-rich diet [5].

Changes in plasma lipid fatty acid levels became more pronounced with duration of ingestion. In the study mentioned above [8], the

TABLE 3.8 Fatty Acid Content in Human Plasma Phospholipids Before, During, and After a Daily Supplement of Cod Liver Oil

Fatty Acid	Fatty Acid Content (Mole %)				
	Before CLO	1 week CLO	2 weeks CLO	1 week after CLO	2 weeks after CLO
LA	22.6± 1.74	19.1± 1.23*	17.8± 1.80*	21.5± 1.42	22.0± 1.26
AA	8.4± 0.65	8.4± 0.70	7.8± 0.90	6.7± 0.48*	7.6± 0.35
EPA	0.8± 0.27	3.1± 0.50**	3.8± 0.64**	1.2± 0.33**	0.7± 0.20
DHA	2.1± 0.43	4.1± 0.86**	4.4± 1.18*	4.0± 0.78*	3.1± 0.64

Mean ± SD (n=6)

* $p< 0.01$ ** $p< 0.001$ compared to values before CLO, Student's t-test

Source: Ref. 14.

TABLE 3.9 Human Plasma Fatty Acid Content Before and During Supplementation with Sardine Oil Concentrate

Fatty acid	Plasma fatty acid content (%)		
	Before supplementation	4 weeks supplementation	13 weeks supplementation
LA	31.5± 4.9	29.7± 5.4[a]	27.3± 5.3[a]
AA	3.8± 1.1	3.9± 0.7	3.5± 0.7
EPA	1.2± 0.7	4.4± 0.7[b]	5.3± 0.9[bc]
DHA	1.9± 0.6	2.7± 0.6[b]	2.9± 0.6[bd]

Mean ± SD (n=12)

[a] $p< 0.005$ compared to values before supplementation
[b] $p< 0.001$ compared to values before supplementation
[c] $p< 0.005$ compared to values at 4 weeks supplementation
[d] $p< 0.025$ compared to values at 4 weeks supplementaiton

Source: Ref. 27.

relative percentage of EPA increased more than 1,000% compared to the basal value, while LA decreased 74% after mackerel was eaten for three days [8]. In another study, following consumption of sardine oil concentrate for four weeks the percentage of LA in plasma lipids dropped 6%, and DHA and EPA increased 42% and 266%, respectively. After nine more weeks of supplementation, the LA level further decreased and DHA and EPA increased 53% and 342% [27] (Table 3.9). Cod liver oil supplementation also caused a rapid, dose-related increase in the relative percent of DHA and EPA in plasma and plasma phospholipids [53].

A return to normal plasma fatty acid levels was observed soon after the cessation of fatty fish or fish oil consumption. Four weeks after 40 ml of cod liver oil was consumed for 25 days the LA, EPA, and DHA levels in plasma phospholipids were not significantly different from what they had been before the feeding trial [20]. Similar results were noted two weeks after ending a two-week daily feeding with 15 ml of the same supplement [14] (Table 3.8). Eight weeks after 40 ml of cod liver oil had been consumed for 4 weeks, DHA in plasma phospholipids had not returned to control values [53].

In summary, the plasma lipid fatty acid content was rapidly altered by consumption of fatty fish or fish oil. Decreases in the LA content and increases in EPA greater than the corresponding ones in DHA were frequently observed. Changes in the AA content were not reported as often.

Plasma Triglycerides

The consumption of fish oils decreased plasma triglyceride (TG) concentration in many trials. These ranged from a 6% decrease (mmol/l) when 200 to 600 g of fatty fish was consumed every week for three months [6] to a 47% decrease when 280 g of mackerel was eaten daily for two weeks [7]. On a percentage basis, the greatest changes in plasma triglyceride concentrations were observed among hyperlipidemic subjects, whose serum TG levels decreased to 50% and 65% of the original levels after nine to twenty-eight days on a salmon flesh and oil diet [10]; mean reductions of 64% and 79% were observed in hypertriglyceridemic individuals who consumed fish oils for four weeks

[11]. In a 20-week feeding study, serum triglyceride levels responded to cod liver oil supplementation in a dose-dependent manner. Serum TG levels were reduced to 66% of control values at the highest doses [40 ml/day] of cod liver oil [53].

In studies using MaxEPA as the dietary supplement, decreases in plasma triglyceride concentrations were reported more frequently [28,31,32,33,34,35,36,38,40,42,49] than not [30]. Decreases in concentration (mmol/l) ranged from 14% when 5 g was consumed daily for three weeks [32b] to 51% after 20 ml was ingested daily for eight weeks [40]. Young men fed varying amounts of MaxEPA for three-week periods showed plasma triglyceride reductions proportional to the supplement dose. MaxEPA at 5 g/day for three weeks produced a 14% decrease in the concentration (mmol/l); MaxEPA at 10 g/day for the same length of time produced a 23% decrease, and when 20 g/day was consumed the decrease was 32% [32b] (Table 3.10).

When 10 g/day of MaxEPA was consumed for fourteen weeks, the plasma triglyceride level dropped progressively as supplementation continued [39] (Figure 3.1). In two feeding trials the MaxEPA supplements were provided for periods of one year or more. In both studies

TABLE 3.10 Plasma Lipid Concentrations After Three Weeks of MaxEPA

Lipid	Plasma lipid concentrations			(mmol/L)	P
	Control	MaxEPA			
		5g/day	10g/day	20g/day	
Total cholesterol	4.21± 0.04	4.06± 0.15	3.98± 0.16	3.82± 0.14	<0.05
HDL cholesterol	1.15± 0.05	1.15± 0.09	1.19± 0.11	1.51± 0.14	<0.05
Triglycerides	0.57± 0.03	0.49± 0.05	0.44± 0.07	0.39± 0.05	<0.01

Mean ± SEM (n=5)

Source: Ref. 32.

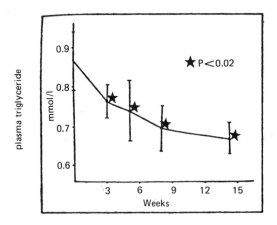

FIGURE 3.1 Influence of a MaxEPA supplement on plasma triglyceride concentrations. N = 12; Mean ± SEM. (From Ref. 39, reprinted by permission from the Medical News Group, Ltd.)

the plasma triglyceride concentrations decreased by approximately 36% after one month of supplementation, and no further decreases were observed during the next eleven [31] and twenty-three months [35] of supplementation (Tables 3.11 and 3.12).

The change in plasma triglyceride level in response to a fat load was monitored in two studies. In one study, the increase after a control meal was compared to that occurring after a salmon-rich meal. Peak levels were seen four hours after the control meal, while after the salmon meal, only slight increases were evident. Significant differences were noted two, four, and six hours postprandially (Figure 3.2). It was postulated that this was due to an accelerated clearance of triglyceride-rich lipoproteins after the salmon meal [9]. Another study compared triglyceride levels following a fat-rich meal before and MaxEPA consumption for four weeks. EPA consumption reduced triglyceride levels and modified the response to dietary fat [35].

Increased levels of EPA and DHA [1,7,14] and either decreased [1] or unchanged levels of LA [7,12,14] were reported for plasma

TABLE 3.11 Changes in Mean Values of Human Serum Lipids Before and During Dietary Supplementation with MaxEPA

Lipid	Plasma lipid concentration (mmol/L)		
	Before	1 month suppl.	12 month suppl.
Triglycerides	2.69	1.71*	1.68*
Cholesterol	6.77	6.56	6.41
HDL Cholesterol	1.22	1.33*	1.20

* $p < 0.001$ (Wilcoxon two sample test); other differences not significant

Source: Ref. 31.

TABLE 3.12 Changes in Human Serum Triglyceride Levels During 24 Months of MaxEPA Supplementation

	Serum triglyceride (mmol/L)		
	Mean	SEM	P
Pre-oil	2.68	0.153	
1 month(n=107)	1.68	0.072	<0.001
3 months(n=91)	1.71	0.019	<0.001
6 months(n=72)	1.57	0.067	<0.001
9 months(n=51)	1.65	0.108	<0.001
12 months(n=24)	1.69	0.108	<0.001
24 months(n=16)	1.58	0.179	<0.001

Source: Ref. 35.

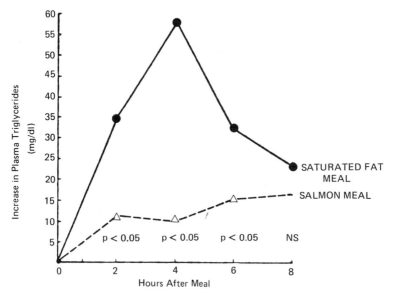

FIGURE 3.2 Fat tolerance tests after the consumption of either a control saturated fat meal or a salmon-based meal. N=10. (From Ref. 9)

triglyceride fatty acids. Contrary to observations made with total plasma lipid fatty acids, the level of AA was increased [7].

In brief, decreases in the concentration of plasma triglyceride were reported in many feeding trials. Fish oil diets had the greatest effect on plasma triglyceride concentrations in individuals with abnormal plasma lipid levels. There was a dose-dependent response of plasma triglyceride levels to MaxEPA consumption and cod liver oil supplementation and, in the feeding trials of longest duration, a leveling off of the response after a month of supplementation. The consumption of salmon or MaxEPA altered plasma triglyceride levels in response to a fat-rich meal.

Plasma Total Cholesterol

The effect of dietary fish oils or n-3 PUFAs on the plasma cholesterol levels of volunteers depended on both the kind of dietary treatment provided and the quantity consumed. No significant change in plasma cholesterol was found in three fatty fish feeding trials

[2,6,7]. In two of these studies, relatively low amounts of fish or n-3 PUFAs were consumed. In contrast, three other studies [1,5,7,9] showed a significant reduction in the mean concentration of plasma cholesterol. These decreases ranged from 7% when 280 g of mackerel was consumed daily for two weeks [7] to 20% when salmon flesh and salmon oil were ingested each day for four weeks [9]. The largest decreases in concentration, 27% and 48% of the initial level, were observed among hypertriglyceridemic individuals who consumed fish oil diets for four weeks [11].

When cod liver oil or a concentrate prepared from it was used as the dietary supplement, no significant change in plasma cholesterol levels was reported [8,16,17,19,20,22,23,53]. Daily consumption of a sardine oil concentrate for four weeks produced no significant effect on plasma cholesterol [26], but in another study cholesterol decreased after similar amounts of oil concentrate were consumed daily for the same length of time [27].

When 92 g of MaxEPA or 120 g of salmon oil was consumed daily for four weeks, a 23% drop in the concentration of plasma cholesterol was observed [38]. Smaller quantities of MaxEPA produced no significant change [28,30,31,32,33,34,39,40] except in a study in which participants were fed 5, 10, and 20 g of MaxEPA daily, in random order, for three-week periods. At the lower dosage, no significant effect was detected, while at the highest dose a mean drop of 9% was observed [32] (Table 3.10).

Two similar studies differed on the effect of purified EPA on plasma cholesterol. One reported no effect [43] and the other reported a significant decrease in concentration [44] after four weeks of supplementation.

Feeding trials using fatty fish [1,7], cod liver oil [14], and MaxEPA [38] showed similar results for cholesterol ester fatty acids: significant decreases in the levels of LA, increases in n-3 PUFA concentrations, and no significant effect on the AA content were observed.

In summary, a significant effect on the plasma cholesterol levels of healthy individuals was observed only in the feeding trials with fatty fish diets and MaxEPA supplementation using the highest dosages. The greatest impact on plasma cholesterol levels was noted among hypertriglyceridemic individuals. A dose-response effect was observed when the same individuals were fed varying amounts of MaxEPA. The effects were maintained only with sustained administration, i.e. a time dependence was consistently observed.

HDL Cholesterol

Changes in HDL cholesterol concentration in response to fatty fish or fish oil consumption were not frequently mentioned. In a feeding trial in which 200 g of mackerel was consumed daily for three weeks, an increase in HDL cholesterol was noted [1]. However, in other feeding trials with fatty fish there was no significant change [5,6,7,9]. A 9% increase in the concentration (mmol/l) of HDL cholesterol was reported in a cod liver oil feeding trial in which 20 ml were fed daily for six weeks [16]; yet no significant changes were observed in three other studies [13,19,53]. Similarly, no increases in HDL cholesterol were noted after supplementation with sardine oil concentrate [26,27]. The consumption of varying amounts of cod liver oil concentrate for four-week periods was not accompanied by a significant concentration change [22,23]. With MaxEPA supplementation, increases in HDL cholesterol concentration were observed as frequently as were no significant changes. When subjects were fed 5, 10, and 20 g/day of MaxEPA in random order for three-week periods, increases in HDL cholesterol were noted only at the highest dose [32b] (Table 3.10).

HDL concentration levels tended to reverse to baseline levels as the supplementation period lengthened. When 10 g of MaxEPA was consumed daily for fourteen weeks, the increase above baseline concentration levels was inversely proportional to the duration of supplementation [39] (Figure 3.3). Similarly, after a month of daily MaxEPA consumption (10 and 20 ml) a 9% increase in the concentration was observed. After another eleven months of supplementation, the concentration had returned to baseline levels [31] (Table 3.11).

Briefly, increases in HDL cholesterol concentration were not frequently observed in feeding trials. With MaxEPA supplementation,

Effects of Omega-3 Polyunsaturated Fatty Acid

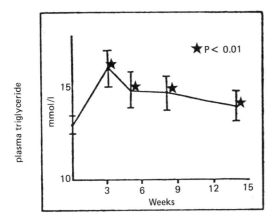

FIGURE 3.3 Influence of a MaxEPA supplement on plasma HDL-cholesterol concentrations. N=12; Mean ± SEM. (From Ref. 39, reprinted by permission from the Medical News Group, Ltd.)

the concentration increased initially and declined as the supplementation period lengthened.

LDL and VLDL Cholesterol

In fatty fish feeding trials, LDL cholesterol concentrations decreased by 15% [5,9]. When 24 g of n-3 PUFAs were consumed daily either as 92 g of MaxEPA or 120 g of salmon oil, a 20% decrease in the level of LDL cholesterol was noted. After kinetic studies of the metabolism of LDL, it was concluded that plasma LDL levels were lowered by a reduction in the rate of synthesis of the apoprotein B component [38].

VLDL cholesterol levels decreased with the consumption of fatty fish [1,5], dropping to about two-thirds of the initial values [5,9]. In a study in which subjects consumed a control diet and then a diet supplemented with MaxEPA, it was concluded that VLDL levels decreased after MaxEPA consumption because of a decrease in the production of apoprotein B [37].

Type V hypertriglyceridemic individuals (with abnormally low LDL levels) showed increases in their mean LDL cholesterol levels after consuming fish oil diets [10,11]. These increases were accompanied by reductions in VLDL cholesterol and VLDL triglyceride levels, suggesting an improved conversion of VLDLs to LDLs [11].

Dietary fish oil may reverse the increases in plasma triglyceride and VLDL levels that accompany high carbohydrate diets. After subjects received such a diet for five days, an 85% increase in plasma triglyceride levels was observed, due mainly to a 126% rise in VLDL triglyceride levels. VLDL cholesterol also rose by 89%. When the fat in the diet was replaced by MaxEPA, the plasma triglyceride levels fell by 61% within three days. This was brought about by a 78% reduction in VLDL triglyceride content and a 65% drop in VLDL cholesterol levels. It was concluded that n-3 PUFAs reduce VLDL levels and plasma triglyceride concentration by inhibiting VLDL triglyceride synthesis [42].

In summary, two separate investigations indicated that the reduced VLDL and LDL levels that follow n-3 PUFA consumption were due to a reduction of apoprotein B synthesis. A decrease in VLDL triglyceride synthesis was suggested as the mechanism by which dietary fish oil lowers the high plasma triglyceride and VLDL levels that accompany a high carbohydrate diet. The consumption of fish oils by Type V hypertriglyceridemic individuals was accompanied by improved VLDL clearance.

Antithrombin III

Antithrombin III reacts with thrombin, Factor Xa, or Factor VIIa to inhibit clotting. The effects of cod liver oil, cod liver oil concentrate, and MaxEPA supplementation on antithrombin III concentration and activity were studied. After the consumption of 20 ml/day of cod liver oil for six weeks, a 16% decrease in the level of antithrombin III was observed. This reduction was maintained five weeks after the supplementation period ended [16]. In another study, after three weeks of cod liver oil concentrate supplementation, a 9% increase in the antithrombin III plasma level was measured, but there was no significant change in activity [24]. Similar results were obtained after 10 g/day of MaxEPA was consumed for four weeks [34].

COMPARISON OF THE FINDINGS FROM FEEDING TRIALS AND THE PLASMATIC CHARACTERISTICS OF ESKIMOS

Some of the plasmatic changes occurring among the participants of feeding trials were comparable to those observed among Eskimos. In plasma lipids, the proportion of n-3 PUFAs increased and the levels of n-6 PUFAs decreased. Reports of reductions in triglyceride levels were numerous. However, decreases in total plasma cholesterol and increases in HDL cholesterol were not observed frequently. Studies of VLDLs and HDLs that showed a lowering in concentration suggested this is caused by a decrease in the synthesis of apoprotein B. Not all studies found increases in the concentration of antithrombin III and not one reported increased antithrombin III activity.

EFFECTS OF FATTY FISH, MARINE OIL, OR n-3 PUFA CONSUMPTION ON PLATELET CHARACTERISTICS

Platelet Lipid Fatty Acids

Changes in the distribution of fatty acids in platelet lipids were reported following consumption of fatty fish diets and dietary supplementation with fish oils, fish oil concentrates, MaxEPA, and purified EPA (Table 3.3). Three studies with fatty fish reported decreased levels of LA and AA and increases in the concentrations of EPA and DHA in platelet phospholipids [2,4,47]. In two cases the percent decreases in AA were smaller than those of LA, and in all, the percent increases in EPA were greater than those of DHA. Three feeding trials with cod liver oil supplementation showed no change in LA level, a decrease in AA, and an increase in EPA greater than that in DHA [8,15,16,20,21,53]. In a study using sardine oil concentrate, there was no change in the AA and DHA levels, but an increase in EPA levels occurred [26]. When MaxEPA was the dietary supplement, no significant change in LA, a decrease in AA, and increases in EPA and DHA were observed [32,32b]. In all types of feeding trials, the increases in EPA were usually 100% or more and greater than the rise in DHA.

Changes in concentrations of platelet lipid fatty acid occurred more slowly than those observed in plasma lipids. Significant increases in EPA and DHA and decreases in LA and AA were observed after six days of a mackerel diet but no changes were noted after only three days [2]. Similarly, a significant increase in EPA was observed after four weeks of consumption of purified EPA but not during or after the initial fortnight [43].

When individuals were fed 5, 10, and 20 g/day of MaxEPA in randomly ordered three-week periods, increases in EPA and DHA and corresponding decreases in AA were observed even at the lowest intake. The trends became more pronounced with increasing dosage [32b] (Table 3.13).

Knapp et al. [55] fed six men with atherosclerosis (angiographically confirmed grade III peripheral vascular disease and showing platelet activation in vivo) 50 ml of MaxEPA fish oil (10 g/day) for 4 weeks as a supplement to a normal diet, which was not specified. There was extensive incorporation of n-3 fatty acids into platelet phospholipids within the first week of feeding. The ratio of n-3 to n-6 PUFAs increased from 0.05 to more than 1.0.

TABLE 3.13 The Influence of Different Doses of MaxEPA Taken for Three-Week Periods on the Fatty Acid Composition of Human Platelet Phosphoglycerides

Fatty acid	Fatty acid composition (weight %)				P
	Control	5g/day	10g/day	20g/day	
LA	7.8± 0.72	8.4± 0.52	8.2± 0.48	7.4± 0.57	N.S
AA	28.9± 0.92	25.2± 1.50	23.0± 0.70	22.7± 0.71	<0.01
EPA	0.9± 0.19	2.3± 0.18	3.2± 0.17	4.2± 0.29	<0.01
DHA	2.4± 0.17	3.1± 0.30	3.3± 0.23	4.3± 0.29	<0.01

Mean ± SEM (n=5)

Source: Ref. 32.

Progressive increases and decreases in platelet lipid fatty acid levels were observed in studies with longer supplementation periods [3,26]. For instance, after three weeks on a fish diet the mean increase in EPA as a percentage of all fatty acids was 200%, after three more weeks the value had increased to 250%, and at eleven weeks it was 300% [3] (Table 3.14).

The EPA and DHA in plasma phospholipids increased after 20 weeks of dietary cod liver oil in parallel with dose and time, with a corresponding decrease in AA [53].

These changes in fatty acid concentration were reversed soon after cessation of supplementation. Thus, five weeks after ending a six-week period of daily cod liver oil supplementation, the AA concentration in platelet lipids returned to normal and EPA and DHA were reduced to one-fifth and one-third of their peak values [15,16].

TABLE 3.14 Changes in Fatty Acid Composition of Human Platelet Phosphatidylcholine Before, During and After a Fatty Fish Diet

Fatty acid	Fatty acid composition (%)					
	Normal Baseline	Fatty fish diet			Post diet	
		3 wks	6 wks	11 wks	6 wks	11 wks
LA	7.1± 0.2	6.1± 0.2[a]	6.0± 0.3[a]	5.9± 0.3[a]	6.7± 0.3	6.5± 0.2
AA	18.3± 0.6	17.5± 0.4	16.8± 0.5	16.9± 0.4	17.9± 0.6	20.0± 0.4
EPA	0.4± 0.1	1.2± 0.1[b]	1.4± 0.2[b]	1.6± 0.2[b]	0.5± 0.1	0.5± 0.1
DHA	1.7± 0.1	2.5± 0.1[b]	2.5± 0.1[b]	2.7± 0.2[b]	1.7± 0.1	1.8± 0.1

Mean ± SEM (n=10)

[a] $p < 0.05$
[b] $p < 0.001$ compared to baseline values

Source: Ref. 3.

In a 20-week cod liver oil supplementation study, DHA decreased more slowly than EPA when the dose was reduced. EPA, DHA, and AA returned to control values 20 weeks after cod liver oil dosing stopped [53]. In another study, two weeks after the end of a two-week period of cod liver oil supplementation, many of the modifications of LA, AA, EPA, and DHA levels in phosphatidylcholine, phosphatidylethanolamine, phosphatidylserine, and phosphatidylinositol were no longer evident [18].

The various classes of platelet phospholipids differed in the type and degree of the change in their fatty acid composition accompanying cod liver oil supplementation [17,18]. After two weeks of cod liver oil, the greatest percent decreases in LA and AA content were observed in phosphatidylserine, while phosphatidylcholine showed the largest percent increases in EPA and DHA. The smallest increases in EPA and DHA were noted in phosphatidylinositol, in which LA and AA content did not significantly change [18]. In another trial in which 25 ml/day of cod liver oil was consumed for six weeks, the greatest mean increase in EPA occurred in phosphatidylethanolamine, and the largest rise in DHA content was in phosphatidylserine and phosphatidylinositol. Only phosphatidylcholine showed a decrease in AA while none of the four classes of phospholipids showed any significant change in LA content [17].

To recapitulate, the consumption of fish or fish oils significantly altered platelet fatty acid concentrations. These changes did not occur as soon after the initiation of supplementation as did modifications in plasma lipid fatty acids. EPA increased severalfold greater than DHA. LA and AA either decreased or did not change significantly in concentration. The fatty acid content of platelet lipids became increasingly modified with duration of supplementation periods. A return to basal values occurred soon after discontinuing supplementation. The EPA levels returned more quickly to control values compared to DHA [53]. The various classes of platelet phospholipids differed in the type and degree of change in fatty acid composition accompanying fish oil supplementation.

Platelet Count

Dietary fatty fish, cod liver oil, sardine oil concentrate, and MaxEPA affected platelet count. Decreases ranging from 8%, when 10 and 20 ml of MaxEPA were consumed daily for one month [31], to 34%, when a salmon-rich diet was consumed for a similar period [4], were observed. In two studies the effect of long-term MaxEPA supplementation on platelet count was monitored [31,35]. The greatest alterations were observed at the beginning of the supplementation period. After one month of daily ingestion of 20 ml of MaxEPA, platelet count was reduced by 9%, and significant reductions continued during the first six months of supplementation. However, in the following 18 months baseline levels were reattained [35]. Thus, it appeared that decreases in platelet count were only temporary.

Platelet numbers of six atherosclerotic patients who received MaxEPA (10 g/day) for four weeks remained reasonably constant; there was a slight shift to a greater number of larger and more serial platelets during supplementation [55].

Platelet Aggregation

Platelet aggregation was studied by adding various concentrations of aggregating substances to platelet-rich plasma (PRP). The aggregating agents most commonly used were AA, adrenaline, ADP, collagen, thrombin, and compound U46619. The percent change in maximum light transmittance or optical density was used as an index of platelet aggregation. There was no significant change from baseline levels in platelet aggregation induced by AA after dietary supplementation or an experimental diet [2,4,8,17,20,11,43,45,46].

Platelet aggregation induced by adrenaline was not altered significantly by diets rich in fatty fish [2,3,4]. With purified EPA consumption, significant reductions in platelet aggregation were observed [43,44,45,48].

Platelet aggregation in the presence of ADP and collagen was studied in PRP from subjects receiving fatty fish, cod liver oil, sardine oil concentrate, MaxEPA, and purified EPA. Declines in platelet aggregation were noted as frequently as no significant change. When subjects were fed 150 g of fatty fish daily for six weeks, platelet fatty acid composition and ADP- and collagen-induced platelet aggregation changed several times before, during, and after the experimental period. The aggregating tendencies of platelets failed to follow changes in their fatty acid composition. Aggregability was decreased weeks after the platelet fatty acid composition had returned to normal, making it unlikely that this altered platelet function was caused solely by changes in total fatty acid levels; however, changes in small but physiologically active pools may not have been detected [47]. In agreement with these findings, when 100 mg of EPA in the form of an arginine salt was consumed daily for two months, platelet aggregability in response to adrenaline, ADP, and collagen was decreased. However, no detectable significant changes in plasma or platelet fatty acid levels occurred [48].

In another study where subjects supplemented their regular diet with 20 to 40 ml cod liver oil daily for 20 weeks, collagen-induced platelet aggregation in PRP decreased to lowest values at the maximum dose of cod liver oil and returned to control values 12 weeks after treatment was stopped [53].

Thrombin was used as the aggregating agent in feeding trials using a salmon-rich diet [4] and cod liver oil [17,18]. No significant change was seen after the fish diet, although platelet aggregation decreased after two weeks of cod liver oil supplementation.

The compound U46619, a PGH_2 (prostaglandin endoperoxide) analogue, was used as an aggregating agent in two MaxEPA supplementation trials. In neither case was platelet aggregation significantly affected by the experimental treatment [32b,33].

In brief, a fatty fish diet or dietary supplementation with fish oil was not observed to influence platelet aggregation induced by AA and compound U46619. There was evidence for a decrease in adrenaline and thrombin-induced platelet aggregation. Reductions in ADP and collagen-induced platelet aggregation were frequently noted.

Thromboxane A_2 Production

Thromboxane B_2 is used as an index of thromboxane production, and its formation in PRP stimulated with AA, adrenaline, collagen, and thrombin was measured by radioimmunoassay. Consumption of mackerel [2] or cod liver oil [8,17,20,21] did not significantly modify the amount of TXB_2 produced in the presence of added AA. A mackerel diet did not significantly affect adrenaline-induced TXB_2 formation in PRP [2]. Collagen-induced formation of TXB_2 decreased in cod liver oil feeding trials [8,17,20,21] and in a MaxEPA supplementation study [33].

The release of labeled AA and the formation of labeled TXB_2 in prelabeled platelets decreased after four weeks of sardine oil concentrate supplementation. However, there was no apparent change in the conversion of endogenous AA to TXB_2 [26].

Thrombin-induced TXB_2 formation in platelets labeled with tritiated AA decreased significantly when 150 mg of EPA was consumed daily for a month. This decrease was not accompanied by detectable changes in plasma or platelet fatty acid composition [45,48]. In addition, thrombin-induced TXB_2 formation from AA-labelled platelets was reduced following consumption of EPA (4.5 g/day) for four weeks [43].

In another feeding trial, blood was collected from bleeding cutaneous wounds before, during, and after a six week period of fatty fish consumption. No significant change in the TXB_2 concentration in nonclotted blood was noted even in subjects whose bleeding times were increased most by the experimental diet. In contrast, the TXB_2 levels in blood collected from cutaneous wounds and allowed to clot by incubation at 37°C for one hour were significantly diminished during and up to eight weeks after the fish diet ended.

This discrepancy may be due partly to the fact that in clotting blood all platelets are activated maximally and produce all the TXA_2 that can be formed from the available AA. Partial replacement of AA by the fatty acids from the fish diet would reduce the amount of TXA_2 produced. In nonclotted blood only a small proportion of platelets are activated and only a part of their total TXA_2-forming capacity is used. This explanation, however, only accounts for the reduced TXA_2 formation during the time the platelet fatty acid composition is altered by dietary means. The persistence of the effects after the end of the diet may involve small pools of physiologically important fatty acids or some yet unknown factors [47].

In summary, neither alterations in AA nor adrenaline-induced TXB_2 formation in PRP was observed in feeding trials. Collagen-induced TXB_2 formation was reduced. A study using platelets prelabeled with AA indicated that with collagen stimulation no change occurs in the rate of conversion of AA to TXB_2, but the release of AA is reduced. Reductions in thrombin-induced TXB_2 formation were observed and the analytical evidence suggests the reductions were not related to the gross changes in fatty acid composition of the platelets or plasma lipids. The consumption of fatty fish decreased TXB_2 production in the serum of clotted blood, but the levels in circulating blood did not change.

In a study of atherosclerotic patients receiving MaxEPA (10 g/day for 4 weeks) [55], serum thromboxane synthesis decreased by 44%, from 429 to 242 ng/ml, during supplementation. In the same study, serum TXB_2 of healthy volunteers decreased from 289 to 97 ng/ml and bleeding time increased from 4.9 to 6.9 minutes, but aggregation responses to arachidonic acid were not significantly altered. There was a rapid, significant fall in excretion of TXB_2 metabolites during the first week in patients with atherosclerosis, from 1406 to 840 pg/mg creatinine; excretion continued to decrease to 588 pg/mg creatinine by the end of 28 days, i.e., 58% decrease overall. The greatest proportional decrease occurred within the initial 7 days.

Thromboxane B_3 Production

Thromboxane B_3 (TXB_3) is the stable metabolite of TXA_3, which is synthesized from EPA and exerts a weak proaggregatory effect on platelets. Upon collagen stimulation, platelets of subjects who had ingested 40 ml of cod liver oil for 25 days formed TXA_3 from endogenous EPA, as measured by the level of TXB_3. The concomitant formation of TXA_2, measured as TXB_2, was reduced compared to the levels formed before the experimental period. The study suggested that TXA_3 may be formed in vivo, and along with a reduced formation of TXA_2, may contribute to reductions in platelet aggregation when EPA is consumed [21].

In a feeding study where participants supplemented their regular diet with 20 to 40 ml cod liver oil for 20 weeks, TXB_2 and TXB_3 values were measured in serum of whole clotted blood by gas chromatography/mass spectrophotometry. TXB_2 values decreased to 50% of control values at the maximum cod liver oil dose and had not returned to control values 8 weeks after cod liver oil supplementation ended. TXB_3 was detected by the same method in small quantities in all samples. The authors concluded that the ingestion of cod liver oil reduced the capacity of platelets to form thromboxane and that the response was dose dependent [53] (Figure 3.5).

Bleeding Time

An increase in cutaneous bleeding time (measured in minutes or seconds) was frequently observed following fatty fish or fish oil

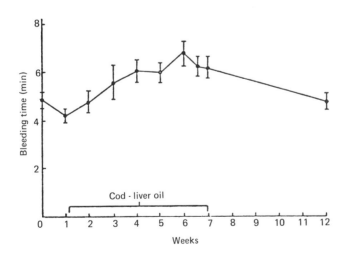

FIGURE 3.4 Change in bleeding time with a cod liver oil supplement. N = 12; Mean ± SEM. (From Ref. 16, reprinted by permission from the authors and the Biochemical Society)

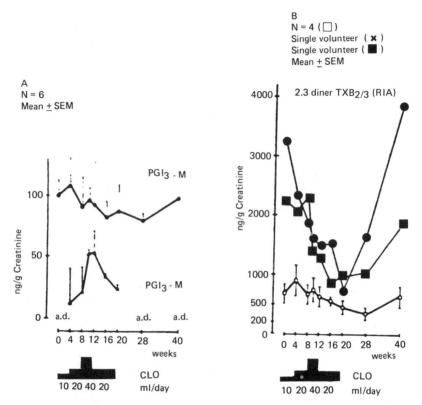

FIGURE 3.5 Endogenous production of prostaglandins I_2 and I_3 (A) and thromboxane $A_{2/3}$ (B) before, during, and after intake of dietary cod liver oil. A. Prostaglandin I_2 and I_3 production was measured by the urinary metabolites prostaglandin I_2-M and I_3-M (ng/g of creatinine) by combined gas chromatography/mass spectrometry. B. Thromboxane $A_{2/3}$ production was measured by the urinary metabolites 2,3-dinor thromboxane $B_{2/3}$ (ng/g of creatinine) by radioimmune assay. (From Ref. 53)

consumption. In two feeding trials with fatty fish, a peak increase of nearly 50% was noted [3,4].

In some studies, bleeding times returned to baseline levels soon after supplementation ceased. Six weeks after ending an eleven week period of abundant fish consumption, the mean bleeding time was not significantly different from its initial value [3], and five weeks after consuming 20 ml/day of cod liver oil for six weeks, bleeding times returned to normal baseline levels [15,16] (Figure 3.4).

Template bleeding times of six atherosclerotic patients receiving MaxEPA supplementation (10 g/day for four weeks) were significantly prolonged in all patients by the end of the first week and remained so for the duration of supplementation, after which they returned to normal levels [55].

When individuals ate 150 g of fatty fish each day, significant changes in platelet fatty acid composition were observed after a week, while significant increases in bleeding time were not detectable until six weeks after the initiation of the feeding trial. There were considerable differences in prolongation of bleeding time between individuals (range 2% to 93%) despite small differences in the fatty acid composition of platelets. Moreover, the mean bleeding time was still significantly prolonged three weeks after the end of the diet when the platelet fatty acid levels had, to a large extent, returned to baseline levels. From these results it was concluded that bleeding time was not related to platelet fatty acid composition [47]. This possibility is supported by another feeding trial which did not show a dose-related response of bleeding time to increasing doses of MaxEPA [32b]. However, this may reflect the fact that subtle changes in the fatty acids in specific pools are occurring but are not detectable by analytical methods.

Briefly, an increased template bleeding time was observed frequently after fish or fish oil consumption. Most studies found that values returned to baseline levels soon after supplementation ceased. Based on total fatty acid patterns, it was suggested that bleeding time is unrelated to changes in platelet fatty acid composition; however, more detailed studies are needed.

Malondialdehyde Production

Malondialdehyde (MDA) is a measure of TXA_2 synthesis. Measurements of its level in PRP were conducted in some feeding trials. Following n-ethylmaleimide (a sulfhydryl blocking agent) administration, the quantity of MDA produced in PRP was 35% greater after a period of abundant salmon intake than after consuming a control diet [4]. MDA production in the presence of AA was also significantly greater after six weeks of cod liver oil supplementation [17]. In two studies with cod liver oil, neither changes in collagen-induced MDA production in PRP nor thrombin-induced MDA production was observed [17,19].

COMPARISON OF THE FINDINGS FROM FEEDING TRIALS AND THE PLATELET CHARACTERISTICS OF ESKIMOS

The changes in LA, AA, EPA, and DHA concentrations in platelet lipids that occured in feeding trials are comparable to those observed among Eskimos. Decreased ADP- and collagen-induced aggregation and increased template bleeding times that were observed frequently are also consistent.

EFFECTS OF FATTY FISH, MARINE OIL, OR n-3 PUFA CONSUMPTION ON THE PRODUCTION OF PROSTAGLANDINS I_2 AND I_3 BY VESSEL WALLS AND ON LEVELS OF URINARY PROSTAGLANDIN METABOLITES

Prostaglandin I_2 (Prostacyclin) Production

The production of prostaglandin I_2 (PGI_2) or prostacyclin by vessel walls was determined using biopsies from forearm veins (Table 3.4). The blood vessel pieces were incubated in a buffer and the level of PGI_2 was determined by the ability of the incubation fluid to inhibit ADP-induced platelet aggregation. Spontaneous release of antiaggregatory substances was not found either before or after six weeks of cod liver oil supplementation. The fluid was also analyzed for 6-keto-prostaglandin $F_1\alpha$, the stable metabolite of PGI_2, after incubation with AA, and none was detected [17].

Prostaglandin I_3 Production

Among young men consuming mackerel, Δ17-2,3-dinor-6-keto-PGF_1, a urinary metabolite of PGI_3, became evident at a level about one-half of that measured for the corresponding PGI_2 metabolite [18]. Studies with smooth muscle cells isolated from human left gastric arteries suggested that human vessel wall cells can produce PGI_3. These cells converted labeled EPA to PGI_3 at levels one-third as great as that of the conversion of AA to PGI_2 [51].

In a recent study [53] in which subjects supplemented their regular diet with 20 to 40 ml cod liver oil daily for 20 weeks, PGI_2 and PGI_3 production was determined by measuring the excretion of prostaglandin urinary metabolites using gas chromatography/mass spectrophotometry. The major urinary metabolite of PGI_3 was detected in urine only during cod liver oil intake. Urinary excretion of the major PGI_2 metabolite did not change significantly during the study. The endogenous production of PGI_3 from ingested EPA increased at rates up to 50% of the PGI_2 production in a dose-dependent manner. These changes lasted throughout the 20-week supplementation period and reverted to control values within 12 weeks after cod liver oil treatment ceased.

Combining these results with the concomitant decrease noted in TXB_2 production in the same study, the authors concluded that cod liver oil supplementation showed a favorable shift of the prostaglandin I/thromboxane A balance to a more pronounced antiaggregatory state [53] (Figure 3.5), which would be consistent with a reduced thrombotic tendency.

When six atherosclerotic patients received MaxEPA (10 g/day for four weeks) supplementation [55], the excretion of prostacyclin metabolite, which had been significantly elevated, decreased during the first week of supplementation from 364 to 269 pg/mg creatinine. This initial decrease of 26% continued throughout the supplementation period, amounting to an approximate 42% decrease during the 28-day feeding period. The final excretion rate attained (approximately 200 pg/mg creatinine) concurred with the excretion rate of healthy volunteers in the study.

In the same study, the excretion rate of PGI_3 metabolite was significantly increased during feeding of fish oil from a baseline level of 7 to approximately 30 pg/mg creatinine. Significantly, fish oil did not decrease excretion of prostacyclin metabolites in healthy volunteers who averaged excretion rates of 100 pg/mg creatinine, and the amount of PGI_3 formed was similar in both groups. The excretion of thromboxane was depressed 50%, from 300 to 147 pg/mg creatinine, in healthy volunteers consuming 50 g fish oil/day.

When atherosclerotic patients were continued on fish oil at a dose of 1 g/day for 6 months after termination of the high dose (10 g MaxEPA/day) both the lipid composition and PG excretion patterns returned to the elevated pretreatment levels [55].

This study showed that very high levels of dietary fish oil significantly decreased serum thromboxane levels and excretion of both thromboxane and prostacyclin but did not completely eliminate the synthesis of thromboxane. These effects coincided with an increase in the EPA concentration in tissues. The lack of depression of prostacyclin in healthy volunteers and its reduction to normal levels of excretion in atherosclerotic patients was noteworthy. This study, like that of von Schacky et al. [53], underscores the need for continued ingestion of very large quantities of fish oil to consistently depress thromboxane synthesis.

EFFECTS OF FATTY FISH, MARINE OIL, OR n-3 PUFA CONSUMPTION ON ERYTHROCYTE CHARACTERISTICS

Erythrocyte Lipid Fatty Acids

After dietary supplementation with cod liver oil and purified EPA, changes in the levels of AA, LA, EPA, and DHA in erythrocyte fatty acids were observed (Table 3.5). The increase in percentage of EPA was often greater than that of DHA [14,15,16,53]. LA and AA concentrations decreased slightly or negligibly [8,14,15,16,20,21,43,53].

The effects of dietary cod liver oil on the fatty acid composition of erythrocyte membrane phosphatidylcholine, phosphatidylethanolamine, phosphatidylserine, and sphingomyelin were compared. Phosphatidylcholine showed the most rapid and greatest alterations in fatty acid composition. After one week of supplementation, its EPA content increased 275%, the DHA level increased by 33%, and the mean AA level was reduced by 7%. During the first two weeks after discontinuing cod liver oil supplementation, the alterations in LA, EPA, and DHA levels in phosphatidylcholine tended to reverse. The LA content returned to basal levels and those of EPA and DHA, while still significantly elevated compared to the initial values, decreased (Table 3.15). The fatty acid composition of phosphatidylethanolamine changed at a slower rate and to a lesser degree than that of phosphatidylcholine. Significant increases in the mean EPA and DHA contents and a decrease in LA were observed only after two weeks of supplementation. The influence of cod liver oil intake on the fatty acid composition of phosphatidylserine was even less. No changes in LA or AA content were observed. An increase in the EPA content was noted after two weeks of supplementation, but a significant rise in DHA was not observed until two weeks after the end of supplementation [14].

In the 20-week supplementation study with cod liver oil, EPA and DHA increased in erythrocyte membrane phospholipids up to 20 weeks,

TABLE 3.15 Changes in Fatty Acid Composition of Human Erythrocyte Phosphatidylcholine Before, During and After a Daily Supplement of Cod Liver Oil

Fatty acid	Fatty acid composition (mole %)				
	Before CLO	1 week CLO	2 weeks CLO	1 week after CLO	2 weeks after CLO
LA	22.4± 1.41	20.9± 1.27[c]	19.5± 1.66[c]	21.7± 1.37[a]	22.5± 1.24
AA	4.9± 0.50	5.2± 0.60	4.9± 0.51	4.6± 0.39	4.6± 0.35
EPA	0.4± 0.17	1.5± 0.66[c]	2.2± 0.37[c]	0.9± 0.16[c]	0.6± 0.22[b]
DHA	1.2± 0.32	1.6± 0.62[b]	2.4± 0.49[c]	2.1± 0.40[c]	1.9± 0.31[b]

Mean ± SD (n=6)

[a] $p<0.05$
[b] $p<0.01$
[c] $p<0.001$, compared to values before CLO Student's paired t-test

Source: Ref. 14.

even after the cod liver oil dose was decreased from 40 ml/day to 20 ml/day at 12 weeks. DHA content did not return to control values 20 weeks after cod liver oil supplementation ended [53].

In the study of six atherosclerotic patients receiving MaxEPA supplementation (10 g/day for 4 weeks) [55], there was extensive incorporation of n-3 fatty acids into erythrocyte phospholipids within the first week of feeding. The ratio of n-3 to n-6 PUFAs increased from 0.05 to greater than 1.0.

In summary, erythrocyte lipid fatty acids levels were altered in response to fish or fish oil consumption. Increases in EPA were much greater than those of DHA. LA and AA showed slight decreases or no significant change in concentration. The fatty acid composition of the various classes of erythrocyte membrane phospholipids changed at different rates and in varying degrees in response to dietary supplementation with cod liver oil. The alterations in erythrocyte lipid fatty acid composition diminished gradually and persisted for some weeks following the end of supplementation.

EFFECTS OF FATTY FISH, MARINE OIL OR n-3 PUFA CONSUMPTION ON OTHER BLOOD CHARACTERISTICS

Erythrocyte Count

No significant change in erythrocyte count was observed in two cod liver oil feeding trials [6,13,14,18,19].

Hematocrit

Hematocrit levels did not change in cod liver oil [8,17,19,20,21] and purified EPA supplementation trials [43].

Hemoglobin

No significant differences were observed between the initial hemoglobin levels and those measured after cod liver oil [8,20,21], cod liver oil concentrate [22,23], or MaxEPA supplementation [30,34,36].

Whole Blood and Plasma Viscosity

Increased whole blood and plasma viscosities have been associated with cardiovascular disease [27]. They were studied in subjects consuming sardine oil concentrate, MaxEPA, and purified

EPA. Blood viscosity was significantly reduced after 1.4 g/day of EPA was consumed for four weeks in the form of a sardine oil concentrate [25]. However, in another study in which a slightly greater amount was consumed daily for thirteen weeks, no significant change was found [27]. Decreases in blood viscosity were also found in a double-blind randomized study using 10 g/day of MaxEPA as a dietary supplement for seven weeks [36]. A similar finding was reported after 4.5 g of EPA ethylester was consumed for four weeks. No significant change in plasma viscosity was reported during any of the feeding trials [27,36,43].

In the study of Knapp et al. [55] there were no apparent adverse effects of fish oil consumption on blood chemistry or clotting functions of six atherosclerotic men. Blood flow was not altered and serum alpha-tocopherol increased from 8.5 to 10.9 ug/ml.

In summary, there was evidence that whole blood, but not plasma, viscosity decreased following ingestion of fish oil.

Blood Pressure

Blood pressure was monitored among participants in fatty fish and fish oil feeding trials. The consumption of 280 g/day of mackerel for two weeks was associated with a mean systolic pressure reduction of 12% and a 9% decrease in diastolic pressure [7]. The same quantity and period of supplementation in subjects with mild essential hypertension decreased systolic pressure by 8% but did not significantly affect diastolic pressure [12]. No significant change was noted when the same quantity of herring was fed over the same period to the same healthy individuals [7]. Decreases in systolic and diastolic pressures of 10% and 15%, respectively, occurred after six weeks of cod liver oil supplementation [15,16]. Supplementation with a sardine oil concentrate caused a mean decrease (9%) in diastolic pressure at the twelfth week of supplementation [27]. A significant decrease in systolic but not diastolic pressure was noted in a double-blind crossover study that used 10 g/day MaxEPA for four weeks [34]. No blood pressure changes were observed in a supplementation trial with purified EPA [43]. Two studies reported that blood pressure levels remained at the lower values several weeks and months after the supplementation.

In brief, blood pressure decreased following consumption of fatty fish or fish oil. In some studies, however, blood pressure levels remained low beyond the end of the supplementation period, but in general, the data were inconsistent.

DISCUSSION

The responses of various components of the vascular system to consumption of fatty fish, marine oils, or purified EPA have been detailed. Certain parameters were frequently altered to resemble the unique characteristics of Eskimos. However, there are serious shortcomings in many of the feeding trials reviewed.

Experimental Design

The experimental design used in many of the feeding trials did not include control nor comparison groups. Without control groups, it is impossible to know the extent to which extraneous factors (dietary history, seasonal variations in blood components, genetics, age, sex, etc.) influenced the findings.

Some studies included a control group, but individuals were not assigned randomly to the control or the treatment group. When two or more measurements were taken on the same groups of individuals (as was done to observe changes with fish or fish oil consumption), the mean value for each of the groups moved toward the total sample mean. The random assignment of individuals of the same population to either the treatment or control groups would have minimized this problem because the groups would have exhibited more similar initial measurements. In the situations where groups with unequal initial measurements were used, analyses of covariance could have controlled the bias. This statistical technique allows one to compare final group means, adjusting for differences in the means of initial measurements.

Because the feeding trial participants were often aware of the treatment they were receiving, the influence of a placebo effect on their responses cannot be ascertained. Furthermore, since the investigators frequently knew which participants were receiving the treatment and which were not, it is also impossible to know how often their expectations influenced their observations.

Effects of Omega-3 Polyunsaturated Fatty Acid

Choice of Subjects and their Compliance with Treatment

Many feeding trials were carried out under the assumption that the subjects' normal diet contained negligible amounts of n-3 PUFAs. Yet only a few studies attempted to determine whether this was true. The methods for doing so included dietary questionnaires [6,17,19,26,41] and seven-day weighted food intake records [32,32b]. Without gathering this information, the screening process for participants cannot be considered complete.

Procedures used to ascertain the subjects' compliance with the experimental treatment included having participants keep a food diary [1,6,22,37], examining for the presence of n-3 PUFAs in plasma lipids [5,38], monitoring the fatty acid composition of platelet membrane phospholipids [36] and erythrocyte phosphoglycerides [39], or weighing the amount of fish oil left in the issued bottles [16,29]. Yet many other trials did not state the method used to monitor compliance with the treatment. It would have been particularly easy for subjects consuming their usual home diet to consume improper amounts of supplements. Among trials that included repeated measurements, it was common to see significant changes during the first days or weeks and a return to baseline levels as supplementation continued. How much this was due to an adaptation process or to a falloff in compliance remains questionable and emphasizes the need for tightly controlled studies.

Side Effects of Fish Oil Consumption

Before increased consumption of fish oils can be recommended, their potentially deleterious effects should be considered. These could result from oxidation and polymerization, a vitamin E deficiency, or vitamin A or D excesses. Steps taken to prevent these problems were described in some reports (Chapter 8). Methods used to stabilize the oils included storing them at low temperatures [22], packaging them in gelatin capsules [22,25,26,32b,43], and adding antioxidants such as dodecyl gallate [32,32b,33] and α-tocopherol [27,37,43]. Some reports included the levels of vitamins A and D contained in the fish oil supplement. The amounts were especially high in cod liver oil. One of the advantages of MaxEPA is its low content of vitamins A and D [33,37].

Other Fish Oil Components

A tendency to overemphasize EPA, disregarding other n-3 PUFAs and compounds present in the fish diet and supplements, also was noted. Very little research has been done on the effects of DHA consumption alone, and only small efforts have been made to separate the effects of the addition of fish to the participants' diet from the results of excluding other foodstuffs, especially dietary lipids.

Comprehensive Studies

Unfortunately, most of the studies reviewed usually were limited to the effects on one or two blood components or biochemical parameters. Comprehensive studies that concurrently follow the effect of fish or fish oil consumption on all the known relevant plasmatic, platelet, vessel wall, and erythrocyte parameters are needed [53].

n-6 and n-3 PUFAs

Research on the interactions between n-6 and n-3 fatty acids, such as the impact of n-3 PUFAs on the enzymes involved in the AA synthetic pathway and the tissue levels of AA, are needed. Studies that monitor the effects of various levels of n-6 PUFAs on the response to dietary fish oil are also necessary. Because the effects of n-3 PUFAs are greatly affected by the amount and type of dietary fat, studies in which dietary fat and n-6 PUFA intake are controlled at varying n-3 PUFA levels are needed.

The efficacy of n-3 PUFAs reflects their capacity to compete with arachidonic acid in the eicosanoid pathway; therefore, it is quite possible that a reduction in both the amount of dietary fat and the proportion of n-6 dietary fatty acids can accentuate the ameliorating effects of dietary fish oil. This has been observed in a number of animal feeding trials (Kinsella, J.E., B. German, and B. Lokesh, unpublished observations, 1986) and is consistent with the suggestions of many researchers that pathophysiological states, such as tumor growth, atherosclerosis, and arthritis, may reflect an imbalance in fatty acid consumption, resulting in excessive eicosanoid production [56-59]. These suggestions dramatize the need for more systematic research to optimize not only fat intake but also the relative proportions of n-6 and n-3 PUFAs required for optimal eicosanoid homeostasis.

Prostaglandin I_3 Production

Feeding trials and studies using cultured cells suggested that human tissues can easily produce PGI_3 from EPA. The effects of fish oil consumption on the serum levels of PGI_3 and the effects of PGI_3 on thrombotic tendencies need to be further elucidated.

Studies Using Other Fish Species

Only relatively few species of fish or their oils have been used in feeding trials. Interest in underutilized fish species and the wiser use of freshwater and saltwater resources could be promoted by testing various fish species in feeding trials.

A direct comparison of the effects of consumption of two fish species was possible from the results of only one of the studies [18]. Much could be learned from a comparative study in which fish or fish oils from several different species are consumed by the same individuals.

**Consideration of the Frequency
and Duration of Fish Consumption**

In all except one of the studies reviewed [6], treatment was provided daily. In order to determine the frequency of consumption needed to bring about the beneficial changes in blood components and platelet behavior, more studies that provide the fish diet or supplement on a less than daily basis are needed.

Also, the potential side effects of long-term fish and fish oil consumption are still largely unknown. Sound, practical guidelines on fish consumption cannot be developed without more extensive knowledge.

Metabolic Pools

Certain metabolic pools containing AA may be differentially affected by dietary fish oils or n-3 PUFAs, i.e., inositol phosphoglycerides involved in platelet metabolism [54]. Research to ascertain the sensitivity of these pools to n-3 PUFA intake, their turnover rate, and the relative rates of AA depletion and replenishment is warranted.

n-3 Polyunsaturated Fatty Acid Dosage

The quantity of fish consumed and the resulting n-3 fatty acid intake in most of the feeding trials was far in excess of what the American population could realistically consume, as summarized in Chapter 6. The observations from clinical feeding studies underscore the problem of exploiting the beneficial effects of fish oil; that is, the required dosage is extremely difficult to achieve in a normal diet. However, the effectiveness of fish oil can be enhanced by reducing the amount of fat consumed and particularly by altering the n-6 PUFA content of that fat.

Studies of the effects of fatty fish consumption at attainable dietary levels are needed. If beneficial cardiovascular changes are observed at those levels, there would be reason to promote an increase in fatty fish consumption in nutrition education programs.

Possible Genetic Determination of Plasma Fatty Acid Levels

Recently, a study suggested that plasma levels of n-3 and n-6 fatty acids may not be determined entirely by dietary intake. After studying Canadians on a western diet and pure and mixed races of Vancouver Island Indians on a diet rich in salmon, investigators proposed that a high plasma EPA content is due to abundant fish consumption but that low levels of AA may be genetically determined [52]. These conclusions are limited by the small number of Indians included in the study. Further investigations of this community and others where seafood is a major dietary component are needed to assess the relative importance of dietary intake and genetic background in determining n-3 and n-6 fatty acid levels and the incidence of cardiovascular disease.

The above research efforts could lead to information that would benefit professionals in the health care field and the public at large.

REFERENCES

1. von Lossonczy, T.O., A. Ruiter, H.C. Bronsgeest-Schoute, C.M. van Gent, and R.J.J. Hermus. (1978). The effect of a fish diet on serum lipids in healthy human subjects. Amer. J. Clin. Nutr. 31:1340-1346.

2. Siess, W., B. Scherer, B. Bohlig, P. Roth, I. Kurzmann, P.C. Weber. (1980). Platelet-membrane fatty acids, platelet aggregation, and thromboxane formation during a mackerel diet. Lancet i:441-444.

3. Thorngren, M. and A. Gustafson. (1981). Effects of 11-week increase in dietary eicosapentaenoic acid on bleeding time, lipids and platelet aggregation. Lancet ii:1190-1193.

4. Goodnight, S.H., W.S. Harris and W.E. Connor. (1981). The effects of dietary w-3 fatty acids on platelet composition and function in man: a prospective study. Blood 58(5):880-885.

5. Harris, W.S., W.E. Connor and M.P. McMurry. (1983). The comparative reductions of the plasma lipids and lipoproteins by dietary polyunsaturated fats: salmon oil versus vegetable oils. Metabolism 32(2):179-184.

6. Fehily, A.M., M.L. Burr, K.M. Phillips and N.M. Deadman. (1983). The effect of fatty fish on plasma lipid and lipoprotein concentrations. Amer. J. Clin. Nutr. 38:349-351.

7. Singer, P., W. Jaeger, M. Wirth, S. Voigt, E. Naumann, S. Zimontkowski, I. Hadju and W. Goedicke. (1983). Lipid and blood pressure-lowering effect of mackerel diet in man. Atherosclerosis 49:99-108.

8. Fischer, S. and P.C. Weber. (1984). Prostaglandin I_3 is formed in vivo in man after dietary eicosapentaenoic acid. Nature 307: 165-168.

9. Harris, W.S. and W.E. Connor. (1980). The effects of salmon oil upon plasma lipids, lipoproteins and triglyceride clearance. Trans. Assoc. Am. Physicians 93:148-155.

10. Phillipson, B.E., W.E. Harris and W.E. Connor. (1981). Reductions of plasma lipids and lipoproteins in hyperlipidemic patients by dietary w-3 fatty acids. Clin. Res. 29(2):628A (abstr).

11. Phillipson, B.E., D.W. Rothrock, W.E. Connor, W.S. Harris, and D.R. Illingworth. (1985). Reduction of plasma lipids, lipoproteins, and apoproteins by dietary fish oils in patients with hypertriglyceridemia. N. Eng. J. Med. 312(19):1210-1216.

12. Singer, P. W. Jaeger, S. Voigt and H. Thiel. (1984). Defective desaturation and elongation of n-6 and n-3 fatty acids hypertensive patients. Prostaglandins, Leukotrienes Med. 15:159-165.

13. Schimke, E., R. Hilderbrandt, O. Beitz, I. Schimke, S. Semmler, G. Honigmann, H.J. Mest, and V. Schliack. (1984). Influence of a cod liver oil diet in diabetics type I on fatty acid patterns and platelet aggregation. Biomed. Biochim. Acta 43 (8/9): 5351-5353.

14. Popp-Snijders, C. J.A. Schouten, A.P. DeJong, E.A. vander Veen. (1984). Effect of dietary cod liver oil on the lipid composition of human erythrocyte membranes. Scand. J. Clin. Lab. Invest. 44:39-46.

15. Sanders, T.A.B., D.J. Naismith, A.P. Haines and M. Vickers. (1980). Cod liver oil, platelet fatty acids and bleeding times. Lancet ii:1189.

16. Sanders, T.A.B., M. Vickers and A.P. Haines. (1981). Effect on blood lipids and haemostasis of a supplement of cod liver oil, rich in eicosapentaenoic and docosahexaenoic acids, in healthy young men. Clin. Sci. 61:317-324.

17. Brox, J.H., J.E. Killie, S. Gunnes and A. Nordoy. (1981). The effect of cod liver oil and corn oil on platelets and vessel walls in man. Thromb. Haemost. 46:601-611

18. Ahmed, A.A. and B.J. Holub. (1984). Alteration and recovery of bleeding times, platelet aggregation and fatty acid composition of individual phospholipids in platelets of human subjects receiving a supplement of cod liver oil. Lipids 19(8):617-624.

19. Brox, J.H., J.E. Killie, B. Osterud, S. Holme, and A. Nordoy. (1984). Effects of cod liver oil on platelets and coagulation in familial hypercholesterolemia (Type IIa). Acta Med. Scand. 213: 137-144.

20. Lorenz, R., U. Spengler, S. Fischer, J. Drum, and P.C. Weber. (1983). Platelet function, thromboxane formation and blood pressure control during supplementation of the Western diet with cod liver oil. Circulation 67(3):504-511.

21. Fischer,S. and P.C. Weber. (1983). Thromboxane A_3 (TXA_3)is formed in human platelets after dietary eicosapentaenoic acid. Biochem. Biophys. Res. Comm. 116(3):1091-1099.

22. Brongeest-Schoute, H.C., C.M. van Gent, J.B. Luten and A. Ruiter. (1981). The effect of various intakes of w-3 fatty acids on the blood lipid composition in healthy human subjects. Amer. J. Clin. Nutr. 34:1752-1757.

23. van Gent, C.M., J.B. Luten, H.C. Brongeest-Schoute and A. Ruiter. (1979). Effect, on serum lipid levels of w-3 fatty acids, of ingesting fish oil concentrate. Lancet ii:1249-1250.

24. Stoffersen, E., K.A. Jorgensen and J. Dyerberg. (1982). Antithrombin III and dietary intake of polyunsaturated fatty acids. Scand. J. Clin. Lab. Invest. 42:83-86.

25. Kobayashi, S. A. Hirai., T. Terano, T. Hamazaki, Y, Yamura, A. Kumagai. (1981). Reduction in blood viscosity by eicosapentaenoic acid. Lancet ii:197.

26. Hirai, A., T.Terano, T. Hamazaki, J, Sajiki, S. Kondo, A. Osawa, T. Fujita, T. Miyamoto, Y. Tamura, and A. Kumagai. (1982). The effects of the oral administration of fish oil concentrate on the release and the metabolism of [^{14}C]arachidonic acid and [^{14}C]eicosapentaenoic acid by human platelets. Thromb. Res. 28: 285-298.

27. Hamazaki. T., R. Nakazawa, S. Tateno, H. Shishido, K. Isoda, Y. Hattori, T. Yoshida, T. Fujita, S. Yano, and A. Kumagai. (1984). Effects of fish oil rich in eicosapentaenoic acid on serum lipid in hyperlipidemic hemodialysis patients. Kidney Int. 26:81-84.

28. Saynor, R. and D. Verel. (1980). Effect of a marine oil high in eicosapentaenoic acid on blood lipids and coagulation. IRCS Med. Sci. 8:378-379.

29. Sanders, T.A.B. and K.M. Younger. (1981). The effect of dietary supplements of w-3 polyunsaturated fatty acids on the fatty acid composition of platelets and plasma choline phosphoglycerides. Br. J. Nutr. 45:613-616.

30. Hay, C.R.M., A.P. Durber and R. Saynor. (1982). Effect of fish oil on platelet kinetics in patients with ischaemic heart disease. Lancet i:1269-1272.

31. Saynor,R., and D. Verel. (1982). Eicosapentaenoic acid, bleeding time, and serum lipids. Lancet ii:272.

32. Sanders, T.A.B. and F. Roshanai. (1983). The influence of different types of w-3 polyunsaturated fatty acids blood lipids and platelet function in healthy volunteers. Clin. Sci. 64: 91-99.

33. Sanders, T.A.B. and M.C. Hochland. (1983). A comparison of the influence on plasma lipids and platelet function of supplements of w-3 and w-6 polyunsaturated fatty acids. Br. J. Nutr. 50:521-529.

34. Mortensen, J.Z., E.B. Schmidt, A.H. Nielsen and J. Dyerberg. (1983). The effect of n-6 and n-3 polyunsaturated fatty acids on hemostasis, blood lipids and blood pressure. Thromb. Haemost. 50:534-546.

35. Saynor, R., D. Verel, and T. Gillott. (1984). The long-term effect of dietary supplementation with fish lipid concentrate on serum lipids, bleeding time, platelets and angina. Atherosclerosis 50:3-10.

36. Woodcock, B.E., E. Smith, W.H. Lambert, W. Morris Jones, J.H. Galloway, M. Greaves and F.E. Preston. (1984). Beneficial effect of fish oil on blood viscosity in peripheral vascular disease. Br. Med. J. 288:592-594.

37. Nestel, P.J., W.E. Connor, M.F. Reardon, S. Connor, S. Wong, and R. Boston. (1984). Suppression by diets rich in fish oil of very low density lipoprotein production in man. J. Clin. Invest. 74:82-89.

38. Illingworth, D.R., W.S. Harris and W.E. Connor. (1984). Inhibition of low density lipoprotein synthesis by dietary omega-3 fatty acids in humans. Arteriosclerosis 4:270-275.

39. Sanders, T.A.B. and M. Mistry. (1984). Controlled trials of fish oil supplements on plasma lipid concentrations. Br. J. Clin. Pract. 38(5):78-81.

40. Bylance, P.B., M.P. George, R. Saynor and M.J. Weston. (1984). A pilot study of the use of MaxEPA in haemodialysis patients. Br. J. Clin. Pract. 38(5):49-52.

41. Harris, W.S., W.E. Connor, and S. Lindsey. (1984). Will dietary w-3 fatty acids change the composition of human milk? Amer.J. Clin. Nutr. 40:780-785.

42. Harris, W.S., W.E. Connor, S.B. Inkeles, and D.R. Illingworth (1984). Dietary omega-3 fatty acids prevent carbohydrate-induced hypertriglyceridemia. Metabolism 33(11): 1016-1019.

43. Terano, T., A. Hirai, T. Hamazaki, S. Kobayashi, T. Fujita, Y. Tamura, and A. Kumagai. (1983). Effect of oral administration of highly purified eicosapentaenoic acid on platelet function, blood viscosity and red cell deformability in healthy human subjects. Atherosclerosis 46:321-331.

44. Nagakawa, Y. H. Orimo, M. Harasawa, I. Morita, K. Yashiro, and S. Murota. (1983). Effect of eicosapentaenoic acid on the platelet aggregation and composition of fatty acid in man. Atherosclerosis 47:71-75.

45. Driss, F., P. Darcet, M. Lagarde, E. Vericel, B. Velardo, M. Guichardent and M. Dechavanne. (1984). Polyunsaturated fatty acids: drug or food? World. Rev. Nutr. Diet. 43:170-173.

46. Velardo, B., M. Lagarde, M. Guichardant, M. Dechavanne, M. Beylot, G. Sautot, and F. Berthezene. (1982). Decrease of platelet activity after intake of small amounts of eicosapentaenoic acid in diabetics. Thromb. Haemost. 48:344.

47. Thorngren M., S. Shafi, and G.V.R. Born. (1984). Delay in primary haemostasis produced by a fish diet without change in local thromboxane A_2. Br. J. Haematol. 58(4);567-578.

48. Driss, F., E. Vericel, M. Lagarde, M. Dechavanne, and P. Darcet. (1984). Inhibition of platelet aggregation and thromboxane synthesis after intake of small amount of eicosapentaenoic acid. Thromb. Res. 36:389-396.

49. Saynor, R. (1984). Effects of w-3 fatty acids on serum lipids. Lancet ii:696-697.

50. Quoted in Discussion, Br. J. Clin. Pract. 38 (5):82, 1984.

51. Morita, I., R. Takahashi, Y. Saito, and S. Murota. (1983). Effects of eicosapentaenoic acid on arachidonic acid metabolism in cultured vascular cells and platelets: species difference. Thromb. Res. 31:211-217.

52. Bates, C., C. van Dam, D.F. Horrobin, N. Morse, Y.S. Huang, and M.S. Manku. (1985). Plasma essential fatty acids in pure and mixed race American Indians on and off a diet exceptionally rich in salmon. Prostaglandins, Leukotrienes Med. 17:77-84.

53. Von Schacky, C., S. Fischer, P.C. Weber. (1985). Lipids, platelet function, and eicosanoid formation in humans. J. Clin. Invest. 76:1626-1631.

54. Simopoulas, A., Kifer, R.R. and Martin, R.E. (1986). Health Effects of Polyunsaturated Fatty Acids in Seafoods. Academic Press, Orlando, Fl.

55. Knapp, H.R., I.A. Reilly, P. Alessandrini, and G.A. Fitzgerald (1986). In vivo indexes of platelet and vascular function during fish oil administration in patients with atherosclerosis. N. Eng. J. Med. 314:937-42.

56. Kinsella, J.E. (1987). Dietary polyunsaturated fatty acids: Effects of n-3 fatty acids in reduction of thrombosis and ischemic heart disease. Nutrition Today (in press).

57. Lands, W.E.M. (1986). Fish and Human Health. Academic Press, New York.

58. Lands, W.E.M. (1986). Renewed questions about polyunsaturated fatty acids. Nutr. Rev. 44:189-95.

59. Dyerberg, J. (1986). Linolenate derived polyunsaturated fatty acids and prevention of atherosclerosis. Nutr. Rev. 44: 126-34.

4
The Effects of Dietary n-3 PUFAs of Fish Oils on Serum Lipids, Eicosanoids, and Thrombotic Events: Observations from Animal Feeding Trials

INTRODUCTION

To elucidate the metabolic mechanisms by which fish consumption ultimately reduces the risk of cardiovascular diseases, numerous feeding trials have been carried out using laboratory animals. A selective literature review of the effects of fish oil consumption in representative mammalian species is presented below.

FEEDING TRIALS

Animal Subjects

The majority of feeding trials reviewed were conducted with weanling to adult male Sprague-Dawley or Wistar rats [4-28]. Other mammalian species occasionally used in dietary studies included guinea pigs [7a], rabbits [29], pigs [30], cats [31], dogs [32], and gerbils [29]. (Table 4.1.)

Marine oils

Several sources of fish oils have been used in the feeding trials: herring [4], menhaden [5,6,7a,7b,8,9,10,11,12,13], sardine [21], and mackerel [30] and cod [14a,14b,15,16,17,18,19,20], squid [22a,22b], and shark liver [6]. MaxEPA, a fish oil concentrate manufactured by British Cod Liver Oils (Hull, England) [24,29], and

TABLE 4.1 A Summary of Marine Oil Feeding Trials Using Rats, Guinea Pigs, Rabbits, Pigs, Cats, Dogs, and Gerbils.

REF.	N	ANIMAL DESCRIPTION	DIETARY TREATMENT	DURATION
I. RATS				
1. HERRING OIL				
(4)	20-40	♂ albino Wistar 6-7 wks; 120-150g	basal diet	10-25 days
	"	"	diets containing: 100g lard/Kg diet	"
	"	"	100g coconut oil/Kg diet	"
	"	"	100g maize oil/Kg diet	"
	"	"	100g herring oil/Kg diet	"
	"	"	100g -irradiated herring oil/Kg diet	"
2. MENHADEN OIL				
(5)	15	♂ weanling Sprague-Dawley	diets containing: 5wt% coconut oil	until aged 36,57,77,100 wks
	"	"	20wt% coconut oil	"
	"	"	5wt% 3:7 corn-menhaden mix	"
	"	"	20wt% 3:7 corn-menhaden mix	"
(6)	10	♂ Sprague-Dawley 250-300g	diets containing: 1.22 wt% safflower oil 3.78 wt% triolein mix	3 weeks
	"			
	"	"	1.22 wt% safflower oil 2.28 wt% triolein mix 1.50 wt % n-3 PUFA enriched menhaden oil	"
	"	"	1.22 wt% safflower oil 2.70 wt% triolein mix 1.09 wt% n-3 PUFA enriched shark liver oil	"
(7b)	5	♂ Sprague-Dawley 400-450g	diets containing: 2 wt% safflower oil 10 wt% hydrog. coconut oil	3 weeks
	6	"	2 wt% safflower oil 5 wt% hydrog. coconut oil 5 wt% menhaden oil	"
(8)	6	♂ Sprague-Dawley 200 g	diets containing: 6.9% soybean oil	3 weeks
	"	"	6.4% menhaden oil/100g	"
(9)	12	♂ Sprague-Dawley 250-300g	diets containing: 10 wt% hydrog. coconut oil	3 weeks

Effects of Dietary n-3 PUFAs of Fish Oils

Table 4.1 (Continued)

REF.	N	ANIMAL DESCRIPTION	DIETARY TREATMENT	DURATION
(9) (cont'd)				
	6	"	5 wt% hydrog. coconut oil 5 wt% menhaden oil	"
	6	"	10 wt% menhaden oil	"
	6	"	20 wt% menhaden oil	"
(10)	4	Sprague-Dawley 400-450g	diets containing: 2 wt% safflower oil 10 wt% hydrog. coconut oil	3 weeks
	4	"	2 wt% safflower oil 10 wt% menhaden oil	"
(11,12)	10	♂ weanling SHR	diets containing: 5% corn oil	22 weeks
	"		1% corn oil, 4% menhaden oil	"
	"	♂ weanling SHR/SP	5% corn oil	"
			1% corn oil, 4% menhaden oil	"
	"	♂ weanling WKY	5% corn oil	"
			1% corn oil, 4% menhaden oil	"
(13)	10	♂ weanling Sprague Dawley; 60-65g	diets containing: 15% hydrogenated coconut oil(HCO)	33 weeks
	"		5% HCO + 10% menhaden oil(ME)	"
	"		5% safflower oil(SAF) + 10% HCO	"
	"		5% SAF + 5% HCO + 5% ME	"
	"		5% SAF + 10% ME	"
3. COD LIVER OIL				
(14a,15)	16	♂ Charles River 175-200g	diets containing: 5 en% corn oil	4 weeks
	"	"	10 en% corn oil	"
	"	"	5 en% corn oil 5 en% cod liver oil	"
(14b,15)	20	♂ Charles River 175-200g	diets containing: 5 en% corn oil	4 weeks
	"	"	5 en% cod liver oil	"
			(both diets administered by gastric intubation)	
(16)	12	♂ Sprague-Dawley 200-250g	diets containing: 5,20,40% safflower oil	4 weeks
	"	"	5,20,40% linseed oil	"
	"	"	5,20,40% coconut oil	"
	"	"	5,20,40% cod liver oil	"

Table 4.1 (Continued)

REF.	N	ANIMAL DESCRIPTION	DIETARY TREATMENT	DURATION
(17)	8	♂ Sprague-Dawley 60-80days; 190-250g	diets containing: 15.0% saturated fat(SF)	6 weeks
	"	"	12.5% SF + 2.5% cod liver oil	"
	"	"	10.0% SF + 5.0% cod liver oil	"
	"	"	12.5% SF + 2.5% linseed oil	"
	"	"	10.0% SF + 5.0% linseed oil	"
(18)	12	♂ Wistar 200-300g	diets containing: 50 en% cod liver oil	8 weeks
	"	"	50 en% sunflowerseed oil	"
(19)	10	♂ Wistar 90g	basal diet: 5wt% fat diets containing:	3,6,12 weeks
	"	"	10 wt% soybean oil	"
	"	"	10 wt% sunflowerseed oil	"
	"	"	10 wt% lard	"
	"	"	10 wt% cod liver oil	"
(20)	NA	♂ Wistar 5 weeks old	diets containing: 5 en% sunflower seed oil	10 weeks
	"	"	50 en% sunflower seed oil	"
	"	"	45 en% codliver oil	"
	"	"	5 en% sunflower seed oil	"

4. SARDINE OIL

(21)	3-6	♂ Wistar 5 weeks old	fat-free diet	2 weeks
	"	"	diets containing: 1% corn oil, 9% HBT	"
	"	"	5% corn oil, 5% HBT	"
	"	"	1% sardine oil, 9% HBT	"
	"	"	5% sardine oil, 5% HBT	"
			(HBT=hydrogenated beef tallow)	

5. SQUID LIVER OIL

(22a)	NA	♂ Sprague-Dawley	diets containing: 2 wt% corn oil	3 weeks
	"	"	3 wt% methyl-oleate	"
	"	"	2 wt% corn oil	"
	"	"	3 wt% PUFA mix from squid liver	
(22b)	6	♂ Sprague-Dawley	diets containing: 5% lard	2 weeks
	"	"	5% methyl-oleate	"
	"	"	5% PUFA mix from squid liver	"
	"	"	2% methyl-oleate + 3% PUFA mix	"
	"	"	4% methyl-oleate + 1% PUFA mix	"
	"	"	5% ethyl-linoleate	"

Table 4.1 (Continued)

REF.	N	ANIMAL DESCRIPTION	DIETARY TREATMENT	DURATION
6. FISH OIL				
(23)	6	♂ Wistar 5 weeks old	fat-free diet	1 month
	"	"	diets containing: 5% corn oil	14 days
	"	"	5% fish oil	14 "
7. MaxEPA				
(24)	NA	♂ hooded Wistar 240-250 g	basal diet	2 weeks
	"	"	basal diet + 15 wt% MaxEPA	"
	"	"	basal diet + 15 wt% safflower oil	"
8. EPA				
(25)	6	♂ Sprague-Dawley 6 weeks old; 170-190 g	diets containing: 6 wt% corn oil	12 days
	"	"	6 wt% corn oil + 0.5 wt% EPA methylester	"
	"	"	6 wt% butter	"
	"	"	6 wt% butter + 0.5 wt% EPA methylester	"
(26)	6	♂ Sprague-Dawley 8 weeks old, 230 g	diets containing: 0.7% linoleic acid	2 weeks
	"	"	0.7% free EPA	"
(27)	NA	♂ Wistar-King 250 g	basal diet	2 weeks
	10	"	diets containing: 100 mg EPA ethylester/day	"
	"	"	100 mg DHA ethylester/day	"
(28)	10	♂ Wistar 7 weeks old	basal diet	8 weeks
	"	"	basal diet + 81 mg EPA ethylester/kg body weight	"
II. GUINEA PIGS				
(7a)	6	guinea pigs	diets containing: 2 wt% safflower oil 10 wt% hydrogenated coconut oil	3 weeks "
	7	"	2 wt% safflower oil 5 wt% hydrogenated coconut oil 5 wt% menhaden oil	3 weeks "
III. RABBITS				
(29)	6	New Zealand rabbits mean initial wt: 1.41 kg	diets containing: 60g corn oil/kg diet	60 days
	"	"	60g linseed oil/kg diet	"
	"	"	60g coconut oil/kg diet	"
	"	"	60g MaxEPA/kg diet	"

Table 4.1 (Continued)

REF.	N	ANIMAL DESCRIPTION	DIETARY TREATMENT	DURATION
IV. PIGS				
(30)	6	♀ Yorkshire piglets 7 weeks old; 15.7 kg. avg. wt.	diets containing: 10% olive oil	4 weeks
	"	"	10% mackerel oil	"
V. CATS				
(31)	5	♂ adult cats, 3-4 kg	basal diet cerebral ischemia induced experimentally with middle cerebral artery occlusion	18-24 days
	"	"	diet containing: 8% calories as menhaden oil cerebral ischemia induced as above	"
VI. DOGS				
(32)	17	♂ mongrel dogs, 12-22 kg	standard chow coronary artery thrombosis induced experimentally by electrical stimulation to left circumflex coronary artery	1-7 days
	10	"	diet containing: 25% calories as menhaden oil coronary artery thrombosis induced as above	36-45 days
VII. GERBILS				
(33)	40	♂ mongolian gerbils, 31-63 g	Mothers fed standard chow during pregnancy; gerbils continued on this diet for 2 months after weaning. Cerebral ischemia induced experimentally with occlusion of right and left common carotid arteries; followed by reperfusion	-
	"	"	Mothers fed diet containing 25% calories as menhaden oil during pregnancy; gerbils continued on this diet for 2 months after weaning. Cerebral ischemia induced as above.	-

Effects of Dietary n-3 PUFAs of Fish Oils

purified EPA and DHA in the form of their ethyl-[27,28] or methylesters [25] have also been fed. The fatty acid composition, source, and method of preparation of the oils is summarized in Table 4.2.

Dosage and Duration of Feeding Trials

Commercial animal feed or laboratory-prepared diets usually made up the major part of the animals' diet. Marine oils were either administered separately or were substituted for other animal fats or vegetable oils in the diet. The marine oil was provided as g/kg of food, as a specified weight or percentage of dietary energy (en%), or in a few cases as a proportion of the animal body weight (wt%).

With rats, the oils were fed for periods ranging from 10 days to 33 weeks. Guinea pigs, rabbits, pigs, cats, and dogs were provided with marine oils for three weeks to two months. The gerbils, whose mothers consumed fish oil during pregnancy, consumed the same diet after weaning.

EFFECT OF MARINE OIL OR n-3 PUFA CONSUMPTION ON PLASMA CHARACTERISTICS

Plasma Lipid Content

The effects of fish oil on numerous parameters were monitored during or after each study and results are summarized in Tables 4.3 to 4.23. Following consumption of fish oils, the plasmatic phospholipid levels in rats were lowered (umol/ml) 16% [21] to 35% [23] compared to those on a fat-free diet for two weeks or one month. When rats were fed either a 1% or 5% sardine oil diet, the decreases in plasmatic phospholipid content were proportional to the intake of dietary fish oil [21], yet no significant differences were observed between rats fed the diets containing sardine oil and those fed the same diets in which the sardine oil was substituted with an equal amount of corn oil [21]. Fish oil consumption did not have the same effect on dogs. No significant difference in plasma lipid content was reported between dogs fed a commercial chow and those consuming 25% of calories as menhaden oil [32]. (Table 4.3)

TABLE 4.2 Fatty Acid Composition of Fish Oils Fed to Laboratory Animals.

Fatty Acid	Menhaden (mole%)	Menhaden (mole%)	Cod Liver (%)	Cod Liver (mole%)	Oil Source Cod Liver (%)	Mackerel (%)	Herring (g/kg ISE)	Squid Liver (%)	Sardine (%)
14:0	9	9.6	4	12.5	7	8.1	6.6±0.1	0.08	-
15:0	-	-	-	-	-	0.6	-	-	-
15:1	-	-	-	-	-	-	-	-	-
16:0	21	20.0	10	27.0	18	14.4	14.4±0.2	0.11	14.3
16:1	11	-	9	12.5	13	5.3	7.2±0.5	1.85	6.9
16:2	-	-	2(+16:3)	-	-	1.1	-	0.09	-
17:1	-	-	-	-	-	-	-	-	-
18:0	3	3.4	2	4.1	-	2.5	2.5±0.04	0.13	-
18:1	12(n9)	11.0(n9)	23	19.5	31	16.2	18.3±0.4	3.04(n9)	14.8(n9)
18:2n6	1	1.3	2	3.1	-	1.6	1.9±0.05	1.09	0.7
18:3	-	-	1	-	-	1.5(n3)	1.3±0.07	1.61(n3)	-
18:4	-	-	-	-	-	4.7(n3)	2.9±0.1	2.16(n6)	-
20:0	-	-	-	-	-	-	-	-	-
20:1	2(n9)	-	13	4.7	12	11.5	9.8±0.4	-	-
20:2n9	-	-	-	-	-	-	-	-	-
20:3n6	-	-	-	-	-	-	-	0.53	-
20:3n9	-	-	-	-	-	-	-	-	-
20:4n3	-	-	-	-	-	-	-	1.93	-
20:4n6	2	2.0(n9)	-	-	-	-	-	1.63	4.0
20:5n3	14	14.1	12	7.7	6	6.3	7.9±0.2	23.68	10.8
21:5n2	-	-	-	-	-	-	-	0.86	-

Effects of Dietary n-3 PUFAs of Fish Oils

	Zapata Haynie (Reedville,VA)	Zapata (Reedville,VA)		Unilever Research Co. (Holland)			Nippon Oil and Fats Co.	Yamakei Sangyo Co. (Japan)	
22:0	–	–	–	–	–	–	–	–	–
22:1	–	–	6(+20:4)	6.1	–	14.1	16.3±0.3	–	–
22:4	–	–	–	–	–	–	–	0.27(n3)	–
22:5n3	2	1.1	2(+22:4)	–	–	0.4	0.7±0.2	6.24	3.6
22:5n6	–	–	–	–	–	–	–	1.06	–
22:6n3	10	8.8	12	2.7	5	8.0	10.1±0.6	37.64	25.6
24:1	–	–	1(+24:1)	–	–	–	–	–	–
24:2	–	–	–	–	–	–	–	–	–
others	–	–	1	–	–	3	–	15.14 / 12.64(artifact)	
Brand Name							Obtained from whole mackerel by cooking and pressing fish mince. Oil was separated in centrifuge, vacuum-dried, treated with activated charcoal and filtered.		
Reference	(31)	(32)	(18)	(14a,14b)	(16)	(30)	(4)	(22)	(21)

TABLE 4.3 Animal Feeding Trials: Summary of Effects of Marine Oil

REF.	DIETS: COMPARED	TO	PLASMA LIPID CONTENT	PLASMA LIPID FATTY ACIDS	TG	CHOL.
I. RATS						
(14a)	5 en% corn oil 5 en% cod liver oil	5 en% corn oil		phospholipids LA　AA　EPA　DHA ↑9%　↓53%　-　- (mole %)		
(14b)	5 en% cod liver oil	5 en% corn oil		phospholipids LA　AA　EPA　DHA ↓51%　↓48%　-　•-(mole %)		
(16)	20,40% cod liver oil	20,40% safflower		phospholipids LA　AA　EPA　DHA NSD　↓　↑　↑(%)		
(21)	1% sardine oil 9% hydrog. beef tallow (HBT)	fat-free diet	phospholipids ↓16% (μmol/ml)	phospholipids LA　AA　EPA　DHA NSD　↓29%　↑　↑99% (%)		
	5% sardine oil 5% HBT	fat-free diet	phospholipids ↓28% (μmol/ml)	phospholipids LA　AA　EPA　DHA ↓44%　NSD　↑　↑154% (%)		
	1% sardine oil 9% HBT	1% corn oil 9% HBT	phospholipids NSD	phospholipids LA　AA　EPA　DHA ↓70%　↓41%　↑　↑77% (%)		
	5% sardine oil 5% HBT	5% corn oil 5% HBT	phospholipids NSD	phospholipids LA　AA　EPA　DHA ↓81%　↓49%　↑　↑288% (%)		
(22a)	3 wt% PUFA mix from squid liver oil	3 wt% methyl-oleate			(mg/100ml) ↓62%	(g/100ml) ↓21%
(22b)	5% PUFA mix from squid liver oil	5% lard			(mg/100ml) ↓43%	
	2% methyl-oleate + 3% PUFA mix	5% lard			(mg/100ml) ↓31%	

Plasma Lipid Fatty Acids

The inclusion of fish oils in the diet produced significant increases in the EPA and DHA levels in plasma lipids of rats and dogs [16,21,23,28,32]. The arachidonic acid (AA) content of phospholipids decreased [14a,14b,21,32] or remained essentially unchanged [21,23,28]. Both increases [14a,23] and decreases [14b,21,32] in the linoleic acid (LA) concentrations in plasma lipids were noted. A comparison of the fatty acid composition of plasmatic phospholipids

Effects of Dietary n-3 PUFAs of Fish Oils and Eicosapentaenoic Acid Consumption on Plasmatic Characteristics.

CHOL. ESTER FATTY ACIDS	HDL-CHOL	LDL-CHOL	VLDL-CHOL	HDL-TG	LDL-TG	VLDL-TG	LIPID PEROXIDE
LA AA EPA DHA ↑54% ↓36% - - (mole %)							
LA AA EPA DHA ↓26% ↓59% ↑ - (mole %)							
		(mg/100ml) ↑109%					(nmol MDA/ml) ↑ 3639%
			(mg/100ml) ↑ 93%				(nmol MDA/ml) ↑ 190%
							(nmol MDA/ml) ↑ 117%

of rats after two weeks on either a fat-free diet or a diet containing variable amounts of sardine or corn oil and hydrogenated beef tallow is shown in Table 4.4 [21]. It was also observed that, irrespective of dietary fat sources, the total proportion of plasmatic unsaturated fatty acids was 53% to 60%. The amount of fatty acids with more than three double bonds was about 18% to 20%. When rats were fed diets containing a limited amount of essential unsaturated fatty acids, the levels of endogenous fatty acids, such as

Table 4.3 (Continued)

REF.	DIETS: COMPARED TO		PLASMA LIPID CONTENT	PLASMA LIPID FATTY ACIDS	TG	CHOL.
I. RATS (continued)						
(22b)	4% methyl-oleate + 1% PUFA mix	5% lard				(mg/100ml) ↓24%
(23)	5% fish oil	fat-free diet	phospholipids ↓35% (μmol/ml)	phophatidylcholine LA AA EPA DHA ↑177% NSD ↑204% ↑ (%)		
(24)	15 wt% MaxEPA	15wt% safflower			(mg/dl) ↓40%	NSD
(27)	100 mg EPA/day	basal diet			NSD	NSD
	100 mg DHA/day	basal diet			NSD	NSD
(27)	basal diet + 81 mg EPA ethylester	basal diet		LA AA EPA DHA ↑ NSD - NSD (μg/ml)	NSD	NSD
II. RABBITS						
(29)	before 60 days consumption of 60g/kg MaxEPA	after 60 days consumption			NSD	NSD
III. PIGS						
(30)	10% mackerel oil	10% olive oil			(mg/100ml) ↓44%	NSD
IV. DOGS						
(32)	25% calories as menhaden oil	standard chow		phospholipids LA AA EPA DHA ↓57% ↓31% ↑471% ↑900% NSD (mole%)		

Comparisons:
- NSD: no significant difference ($p \leq 0.05$) between comparison times
- ↓ : significant decrease between comparison times
- ↑ : significant increase between comparison times

16:1, 18:1, and 20:3n-6 increased. It was concluded that these levels were in some way physiologically necessary and that the desaturation of fatty acids may be regulated by dietary unsaturated fatty acids [21].

Changes in plasmatic lipid fatty acids occurred quickly following introduction of dietary alterations but were rapidly reversed upon removal of fish from the diet. During a feeding trial in which linoleate-deficient rats were fed a fish oil diet for fourteen days and then a fat-free diet for the following fortnight,

Table 4.3 (Continued)

CHOL. ESTER FATTY ACIDS	HDL-CHOL	LDL-CHOL	VLDL-CHOL	HDL-TG	LDL-TG	VLDL-TG	LIPID PEROXIDE
	(mg/100ml) ↑87%						(nmol MDA/ml) ↑168%
	NSD (mg/dl)	NSD	NSD	NSD	NSD (mg/dl)	NSD	
	↑11%	NSD	NSD	NSD	↓44%	NSD	
NSD							

changes in the composition of phospholipid fatty acids over time was followed. Changes were also monitored when the dietary process was reversed. The consumption of a 5% fish oil diet was accompanied by a drop in 20:3n-9 and a rise in n-3 PUFAs. The level of n-3 PUFAs continued to increase during the fourteen days on the fish oil diet. By the fourteenth day, 20:3n-9 had reached very low levels. With the return to a fat-free diet from a diet containing fish oil, the levels of n-3 PUFAs decreased to 45% of their peak

TABLE 4.4 Effect of Feeding Fat-free, Corn Oil and Fish Oil Containing Diets to Rats for 2 Weeks on the Fatty Acid Composition of Plasmatic Phospholipids. Mean ± S.D.; N=3-6.

Dietary Fat		Fatty Acid Content of Plasma Phospholipids (%)			
%	source	LA	AA	EPA	DHA
fat free diet		5.53±0.8	10.7±1.5	--	3.97±0.6
1% sardine oil 9% HBT*		4.41±0.2[b]	7.6±1.5[a,b]	3.46±0.8	7.91±0.3[a,b]
5% sardine oil 5% HBT		3.09±0.6[a,b]	8.11±1.3[b]	8.53±0.1	10.1±0.6[a,b]
1% corn oil 9% HBT		14.7±1.6[a]	12.8±2.8	--	3.91±0.3
5% corn oil 5% HBT		16.0±2.2[a]	15.8±0.7[a]	--	2.60±0.9

*hydrogenated beef tallow

[a]significantly different (p<0.02) from fat-free

[b]significantly different (p<0.02) from same percent of corn oil

Source: Ref. 21.

levels by the sixth day while a concomitant rise in 20:3n-9 was evident [23] (Figure 4.1).

In summary, fish oil consumption caused declines in plasmatic lipid content in rats on a fat-free diet, but not in those on a diet containing vegetable oil. The plasmatic lipid content of dogs was not affected by a 25% fish oil diet. The dietary intake of fish oil was accompanied by rapid increases in the plasma content of EPA and DHA and variable modifications of the LA and AA levels.

Plasma Triglycerides

In the trials in which fish oils constituted a greater proportion of the diet, plasma triglycerides decreased [22a,24,30]. Rats fed a diet containing 15 wt% MaxEPA for two weeks

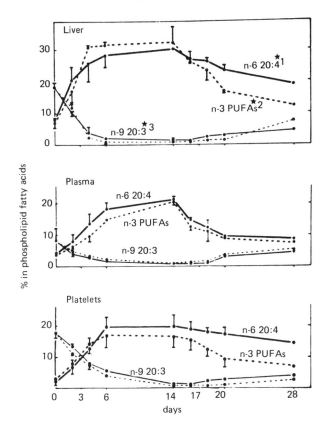

FIGURE 4.1 Effect of dietary fats on polyunsaturated fatty acid levels in rat plasma, platelets, and liver phospholipids. Linoleate-deficient rats were fed a 5% corn oil (._____.) or fish oil (.....) diet for fourteen days and then a fat-free diet for the following fourteen days. In the fish oil group, the proportions of n-3 PUFAs (EPA and DHA) and 20:3n-9 were followed during dietary manipulation. Mean ± S.D.; n=6. (From Ref. 23, reprinted by permission from the authors and Elsevier Biomedical Press.)

had a mean plasma triglyceride level 40% lower than did rats fed the same quantity of safflower oil [24]. After examining the metabolic changes that accompanied the decrease, the investigators concluded that the decrease was due to a diminished rate of lipogenesis, an increase in fatty acid oxidation, and a diminished secretion of triacylglycerides by the liver [24]. When only 81 mg EPA/kg body

TABLE 4.5 Total Plasma Cholesterol and HDL-Cholesterol of Rats Fed Lard or a Polyunsaturated Fatty Acid Mix from Squid Liver for 2 Weeks. Mean ± S.E.; N=6.

Diet	Total Cholesterol (ng/100ml)	HDL-Cholesterol
basal diet +5% lard	192.8 ± 16.6[a]	26.6 ± 2.0[a]
+5% PUFA mix	109.8 ± 7.0[b]	55.7 ± 1.3[b]
+3% PUFA mix 2% methyloleate	132.4 ± 13.3[bc]	51.3 ± 1.3[b]
+1% PUFA mix 4% methyloleate	146.0 ± 8.2[c]	49.7 ± 3.2[b]

Values in same column but not sharing a common superscript letter are significantly different.

Source: Ref. 22.

weight [28] or 100 mg of EPA and DHA [27] were given daily, no significant differences were observed between the mean plasma triglyceride levels of control and treatment rats.

Plasma Cholesterol

The effects of dietary fish oils or n-3 PUFAs on plasma cholesterol levels in laboratory animals were unclear. Feeding rats a squid liver PUFA mix or a combination of the mix and methyl oleate decreased plasma cholesterol levels (mg/100 ml) below the cholesterol levels of rats fed diets containing the same total percentage of fat in the form of lard. The greatest alteration in cholesterol level, a 43% drop, was observed with a 5% PUFA mix diet; mixtures of the PUFA mix and methyl oleate produced changes proportional to their PUFA content [22b] (Table 4.5). However, in other feeding trials in which a 15 wt% MaxEPA diet [24] and 100 mg of EPA and DHA ethylesters [27] were consumed by rats and a 10% mackerel oil diet was consumed by pigs [30], no significant changes in plasma cholesterol levels were observed.

The fatty acid composition of cholesterol esters was altered in rats fed cod liver oil. Decreases in the AA content were observed [14a,14b]. The EPA level was noted to rise [14b], and both an increase [14a] and a decrease [14b] in the LA content were observed in different studies.

In brief, plasma cholesterol decreased in proportion to the n-3 PUFA content of the diet in rats. A decrease in the AA content and a rise in EPA levels in cholesterol esters were observed with fish oil consumption.

HDL Cholesterol

Compared to rats fed a 5% lard diet, groups fed a squid liver PUFA mix with or without methyl oleate had increased mean HDL cholesterol levels (mg/100 ml). Consumption of the PUFA mix without methyl oleate resulted in a 109% increase. Similar increases were observed when part of the PUFA mix was replaced by methyl oleate [22b]. Table 4.5 shows the lack of HDL cholesterol level response to n-3 PUFA intake.

The HDL cholesterol levels, which may be more sensitive to DHA than to EPA intake, increased by a mean of 11% after 100 mg of DHA was given to rats daily for two weeks. No significant change was observed when the same quantity of EPA was given [27]. In addition, no significant difference in HDL cholesterol was noted among rats fed a commercial diet with or without EPA ethylester supplementation [28].

LDL and VLDL Cholesterol

In the only feeding trial which examined LDL and VLDL cholesterol levels, the levels were not significantly modified by either EPA- or DHA-ethylester consumption by rats [27].

In the same feeding trial, consumption of EPA and DHA ethylesters did not significantly change the mean levels of HDL and VLDL triglycerides [27]. In contrast, LDL triglyceride levels (mg/dl) were decreased 44% with DHA ethylester consumption while no effect was observed with EPA consumption [27].

TABLE 4.6 Animal Feeding Trials: Summary of Effects of Marine Oil and Eicosapentaenoic Acid on Platelet Characteristics.

REF.	DIETS COMPARED TO	PLATELET LIPID CONTENT	PLATELET LIPID FATTY ACIDS	PLATELET COUNT	PLATELET AGGREGATION AA	PLATELET AGGREGATION ADP	PLATELET AGGREGATION COLLAGEN	PLATELET AGGREGATION THROMBIN	TXB$_2$ ADP	FORMATION THROMBIN	BLEEDING TIME (SEC)	MDA PRODUCTION COLLAGEN
I. RATS												
(6)	1.5 wt% enriched menhaden oil 3.78 wt% TM 2.28 wt% triolein mix (TM) 1.22 wt% safflower mix (SM)		phospholipids LA AA EPA DHA NSD NSD NSD ↑224 NSD (wt%)		(5x10^{-7}/- 5x10^{-4}) NSD		(.08–.5 mg/ml) NSD				NSD	
	1.09 wt% enriched shark oil 2.70 wt% triolein 3.78 wt% TM mix (TM) 1.22 wt% safflower 1.22 wt% SM mix (SM)		phospholipids LA AA EPA DHA NSD NSD NSD NSD (wt%)		NSD		NSD					NSD
(14a)	5 en% corn oil 5 en% cod liver oil		LA AA EPA DHA ↑32% ↓9% − − (mole%)			NSD				10 U/ml ↓30%		
(14b)	5 en% cod liver oil 5 en% corn oil		LA AA EPA DHA ↓24% ↓39% − − (mole%)			NSD				10 U/ml ↓33%		
(17)	12.5% saturated fat 15% saturated 2.5% cod liver oil fat								(3,5uM) NSD			
	10% saturated fat 15% saturated 5% cod liver oil fat								(3,5uM) NSD			
(18)	50 en% cod liver 5 en% oil sunflower seed oil											
(20)	45 en% cod liver 50 en% sun- 5 en% sunflower seed flower seed oil oil										↑(sec) ↑52%	↓(ng/10^9 platelets)
(23)	5% fish oil fat-free diet	phospholipid phosphatidylcholine NSD LA AA EPA DHA NSD ↑38% ↑ ↑(%)										↓(nmol/10^9 platelets)

I. RATS (continued)
(26) 0.7% free EPA   0.7% linoleic acid   NSD   NSD

(27) 100 mg EPA/day   basal diet   NSD   5uM   15ug/ml
                                                                             ↓60% ↓60%
     100 mg DHA/day   basal diet   NSD   5uM   15ug/ml
                                                                             ↓60% ↓54%

(27) basal diet + basal diet   (10,20, (40,80ug/ml)
     81 mg EPA ethylester   40 uM)
     /Kg body weight   NSD   NSD

II. RABBITS   (compared to corn oil fed rabbits)
(29) before 60 days after 60   LA AA   EPA   DHA   (9.25-74.25uM)(1-25ug/50mL)(2.5U/40ulH₂O)
     consumption of 60g/kg days con- ↓63% ↓58% ↑315%↑   ↓   NSD
     MaxEPA   sumption   (%FA)

III. DOGS   (0.66,2,5 (2,11,50
(32) 25% calories as standard   LA AA EPA   DHA   7,10mM) 110Mu) (3,6ug/ml)
     menhaden oil   chow   ↓51% NSD ↑566% ↑866%   NSD   NSD   NSD
                                       (mole %)

Comparisons:

NSD: no significant difference (p≤0.05) between comparison times

↓ : significant decrease between comparison times

↑ : significant increase between comparison times

Lipid Peroxide

The concentration of lipid peroxide (measured as nmol malondialdehyde/ml) was increased by 3600% in rats fed a diet containing 3 wt% squid liver PUFA mix when compared to rats fed the same quantity of methyl oleate [22a]. Rats fed combinations of the PUFA mix and methyl oleate had similar lipid peroxide levels that were 117% to 190% greater than those of rats fed a diet with the same total percentage of fat as lard [22b]. These increases were accompanied by increases in hepatic lipid peroxide levels (Table 4.18).

EFFECTS OF MARINE OIL OR n-3 PUFA CONSUMPTION ON PLATELET CHARACTERISTICS

Platelet Lipid Content

Because of their importance in hemostasis, vascular integrity, and thrombotic phenomena, the effects of dietary fish oil on platelet functions have been studied. The phospholipid content of rat platelets did not change significantly after rats were fed a fat-free diet for one month or after rats consumed a 5% fish oil diet for fourteen days [23]. (Table 4.6.)

Platelet Lipid Fatty Acids

Fish oil diets produced significant changes in platelet lipid fatty acid composition. Increases in EPA and DHA levels were noted frequently [20,23,29,32]. The effect on LA and AA contents was more variable; both significant increases and decreases were observed following fish oil consumption [14a,14b,23,29,32].

Contrasting effects on platelet lipid fatty acid compositions were observed when rats were fed diets abundant in EPA or DHA (Table 4.7). The source of EPA was an n-3 PUFA enriched menhaden oil, which markedly increased platelet lipid EPA compared to rats fed a control diet. The platelet lipid DHA content was unaltered. The source of DHA was an n-3 PUFA enriched shark liver oil, which caused no significant changes in the EPA or DHA contents of platelets; however, DHA levels were almost twofold higher than levels in the diet containing menhaden oil. It appeared that DHA was not easily incorporated into platelet phospholipids [6].

TABLE 4.7 Fatty Acid Composition of Platelet Phospholipids from Rats Fed EPA and DHA Containing Diets for 3 Weeks. Mean.

Fatty Acid	Control[1] (N=4)	EPA[2] (N=3)	DHA[3]
	Platelet Phospholipid Fatty Acid Content (wt%)		
LA	4.3	5.0	4.0
AA	16.8	10.9	10.9
EPA	1.7[a]	5.5[b]	3.5[ab]
DHA	<1	<1	<1

[1] fat content of diet: 1.22 wt% safflower oil, 3.78 wt% triolein mix
[2] fat content of diet: 1.22 wt% safflower oil, 2.28 wt% triolein mix 1.50 wt% w3-PUFA-enriched menhaden oil (EPA 29.2%)
[3] fat content of diet: 1.22 wt% safflower oil, 2.70 wt% triolein mix, 1.09 wt% w3-PUFA-enriched shark liver oil (DHA 60.1%)

[ab] values having different superscripts are significantly different at $p<.05$

Source: Ref. 6.

In another feeding trial, rats were fed diets containing 6 wt% corn oil or 6 wt% butter with and without 0.5 wt% EPA methylester. Corn oil was used as a source of LA, and butter was chosen because of its low LA content. After twelve days, the fatty acid compositions of platelet phospholipids in the four groups were significantly different. As shown in Table 4.8, rats fed butter experienced a significant reduction in platelet AA content. The ratio of EPA to AA was greater in the butter-fed group than in the corn oil-fed group. The inclusion of 0.5% EPA in the diet markedly increased the ratio. The largest platelet EPA/AA value was observed in the group fed butter and EPA. EPA supplementation caused a significant increase in platelet EPA and a greater decrease in AA. From these observations, it was concluded that the incorporation of EPA into platelet

TABLE 4.8 Fatty Acid Composition of Platelet Phospholipids of Rats Fed Corn Oil and Butter Containing Diets With and Without Eicosapentaenoic Methylester Supplementation for 12 Days. Mean; N=6.

Fatty Acid	Diet			
	Corn Oil	Corn Oil +EPA	Butter	Butter +EPA
	Fatty Acid Content of Phospholipids (%)			
LA	10.7	10.8	5.7	5.1
AA	26.3	19.2	22.1	15.5
EPA	1.0	5.9	1.1	10.9
EPA/AA	0.038	0.036	0.050	0.703

Source: Ref. 25.

phospholipids was enhanced by eliminating LA and AA from the diet [25].

The changes in platelet fatty acids over time was followed in linoleate-deficient rats fed a 5% fish oil diet for fourteen days and then a fat-free diet. The feeding of fish oil caused a rapid substitution of 20:3n-9 by n-3 PUFAs. The concentration of n-3 PUFAs plateaued by the sixth day of feeding, while 20:3n-6 levels greatly diminished. When the rats resumed a fat-free diet, the n-3 PUFAs decreased at a slower rate than they had increased [23].

In brief, EPA and DHA levels in platelet lipids usually increased following fish oil consumption. The effects of fish oil consumption on LA and AA levels were conflicting. Of the n-3 PUFAs, DHA was less easily incorporated into platelet phospholipids than EPA. EPA levels in platelet phospholipids were increased by limiting or eliminating AA and LA from the diet. When followed over time, n-3 PUFAs were incorporated into platelet phospholipids and reached peak levels within a few days; their levels declined more slowly when fish oil consumption ceased.

Platelet Count

The mean platelet count of rats was not significantly affected by the daily consumption of either EPA or DHA ethylester [27].

Platelet Aggregation

The induction of platelet aggregation was measured in platelet--rich plasma (PRP). No signficant difference in AA-induced platelet aggregation was observed among dogs fed only a commercial chow and those whose diet contained 25% of calories as menhaden oil [32].

ADP-stimulated platelet aggregation was not significantly altered by fish oil consumption [6,14a,14b,26,28,32]. The ADP concentration required to give a 50% reduction in platelet aggregation did not change significantly when rats were fed diets containing corn oil or corn oil and cod liver oil [14a]. The same was true when rats were fed 5% corn oil and 5% cod liver oil [14b]. The minimum concentration of PGI_2 that inhibited aggregation did not change significantly in the treatment and control groups [14a,14b].

However, as shown in Table 4.9, platelet aggregation was reduced 60% after rats consumed either 100 mg/day of EPA or DHA for two weeks [27]. In addition, rats fed 60 g/kg MaxEPA for sixty days

TABLE 4.9 Effect of Feeding EPA and DHA Containing Diets for 2 Weeks on Rat Platelet Aggregation. Mean ± S.E.M.; N=10.

Diet	Aggregating Agent	
	Collagen (15 g/ml)	ADP (5 M)
	Percent of Maximum Aggregation	
Basal Diet	31.8 ± 4.0	13.4 ± 2.2
+ 100 mg EPA/day	14.0 ± 2.1*	5.3 ± 0.6 *
+ 100 mg DHA/day	14.8 ± 1.6*	5.3 ± 0.6 *

* statistically significant to basal diet values ($p < 0.0005$)

Source: Ref. 27

had a significantly lower initial rate of ADP-induced platelet aggregation than rabbits fed corn oil. The extent of aggregation was not significantly influenced by diet at levels of ADP below 9.25 μM. At higher concentrations, however, platelet aggregation was significantly greater in rabbits fed corn oil than in those fed fish oil [29].

It often was reported that collagen-induced platelet aggregation was not affected by fish oil consumption [6,26,28,32]. However, PRP from rats fed 100 mg EPA/day for two weeks showed a 60% decrease in aggregation compared to control animals, while those fed 100 mg DHA/day had a 54% drop [27] (Table 4.9). Rabbits fed MaxEPA showed a lower initial rate and extent of aggregation in the presence of collagen than animals fed a corn oil diet [29].

After twelve days of feeding either corn oil or butter to rats, with and without 0.5% EPA methylester supplementation, platelet aggregation in the presence of collagen was compared (Figure 4.2). There was little difference in platelet aggregation between rats consuming corn oil alone and corn oil and EPA. In contrast,

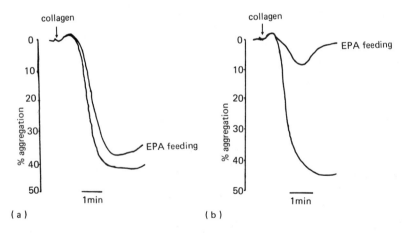

FIGURE 4.2 Effect of EPA supplementation on rat platelet aggregations stimulated by collagen (4.2 ug/ml; n=6). Platelets were isolated from rats on 6 wt% corn oil (A) or 6 wt% butter (B) diets with and without 0.5 wt% EPA supplementation for twelve days. (From Ref. 25, reprinted by permission from Churchill Livingstone.)

markedly less aggregation was observed in the group fed butter and EPA than in the butter-fed group. This difference was attributed to the different amounts of dietary LA and EPA; as the diets became relatively poorer in LA and richer in EPA, the incorporation of EPA into platelet phospholipids increased [25]. Thrombin was used as an aggregating agent in a feeding trial using rabbits. Rabbits that consumed 60 g/kg of corn oil or MaxEPA for sixty days did not show significant differences in platelet aggregation in the presence of thrombin [29].

In summary, platelet aggregation induced by AA and thrombin was not significantly affected by fish oil consumption. Platelet behavior in the presence of ADP and collagen was variable. Differing concentrations of aggregation-inducing agents, quantities of fish oil, and other dietary fats account for the variability. The n-3:n-6 ratio in the diet, and consequently in platelet phospholipids, may also be influential, as was observed in collagen-induced aggregation.

Thromboxane Production

Because of its critical role in platelet function, the amount of TXB_2 produced during platelet aggregation was measured by radioimmunoassay. Using ADP as an inducing agent, TXB_2 production in platelets of rats fed diets containing saturated fat and cod liver oil or only saturated fat was similar [17]. In the presence of thrombin, on the other hand, platelets of rats fed cod liver oil produced about one-third less TXB_2 than did those of animals fed a corn oil diet [14a,14b].

The consumption of fish oils decreased thromboxane levels in the serum of clotted blood. Compared to rats fed a diet containing 5 en% safflower oil, those fed a diet containing 5 en% cod liver oil had a 26% lower level of TXB_2 in the serum of clotted blood. When the oils were consumed as 20 en% and 40 en% of the diet, even greater differences between TXB2 levels in rats fed the two types of oil were seen [16]. In another feeding trial, TXB_2 in serum of clotted blood was depressed by a mean of 48% in rats fed n-3 PUFA enriched menhaden oil (abundant in EPA) but not in those fed n-3 PUFA enriched shark liver oil (more abundant in DHA) [6].

Bleeding Time

Bleeding time (measured as the time required for blood flow from the right saphenous vein through a syringe needle to cease) did not increase significantly in rats fed diets containing either 1.5 wt% n-3 PUFA enriched menhaden oil or 1.09 wt% n-3 PUFA enriched shark liver oil [6]. In contrast, rats fed a diet with 45 en% cod liver oil showed a 52% increase in bleeding time when the tip of the tail was transected and immersed in saline [20]. In addition to the different proportions of fish oil in the diet, the conflicting results also could be related to the different body sites and methods chosen.

Malondialdehyde (MDA) Production

MDA levels correlate with the production of endoperoxides and thromboxane from AA [18,20]. In two feeding trials a supramaximal dose of collagen caused decreased MDA production by platelets (ng or nmol/10^9 platelets) of rats fed diets rich in cod liver oil when compared to MDA synthesis by platelets of corn oil fed rats [18,20].

Production of HHT, HHTE, HETE, and HEPE

HHT [(5Z,8E,10E)-12-L-hydroxyheptadecatrienoic acid] and HHTE [(5Z,8E,10E,14Z)-12-L-hydroxyheptadecatetraenoic acid] are stable cyclooxygenase products of platelet AA and EPA, respectively, and reflect TXA_2 and TXA_3 formation. HETE [(5Z,8Z,10E,14Z)-12-L-hydroxyeicosatetraenoic acid] and HEPE [(5Z,8Z,10E,14Z,17Z)-12-L-hydroxyeicosapentaenoic acid] are the corresponding platelet lipoxygenase products [20].

When platelets of rats fed either 50 en% sunflower seed oil or 45 en% cod liver oil and 5 en% sunflower seed oil were stimulated with varying doses of collagen, the platelets of rats fed fish oil produced 75% less cyclooxygenase products and 50% less lipoxygenase products (Figure 4.3). Using HPLC, it was observed that TXA_3 (measured as HHTE) was not formed in significant amounts, whereas HEPE was 25% of the total lipoxygenase products [20]. These reports suggested that cod liver oil caused a relative deficiency of AA and a decreased formation of prostaglandins of the 2-series. Prostaglandins of the 3-series were formed in very small amounts as EPA was acted upon more readily by lipoxygenase than by cyclooxygenase [20].

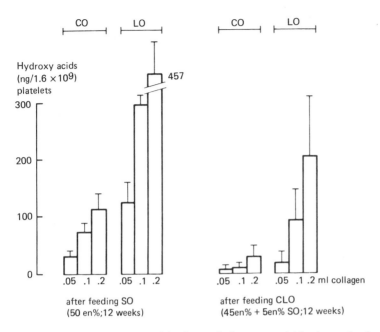

FIGURE 4.3 Hydroxyacids formed from arachidonic and eicosapentaenoic acids by collagen-induced platelets from rats fed diets containing either sunflowerseed oil (SO) or cod liver oil (CO). Mean ± S.D.; CO: cyclooxygenase products HHT and HHTE; LO: lipoxygenase products HETE and HEPE; Wilcoxon two sample test: $p<0.1$ for 50 and 100 ul collagen dose, $p<0.001$ for 200 ul collagen dose. (From Ref. 20, reprinted by permission from Geron-X, Inc. Publishers.)

EFFECTS OF MARINE OIL OR n-3 PUFA CONSUMPTION ON CARDIAC CHARACTERISTICS

Cardiac Lipid Content

The phospholipid contents (mol/ml) of hearts of rats fed a fat-free diet or a diet containing sardine or corn oil and hydrogenated beef tallow were not significantly different [21]. Similarly, the cardiac total lipid (g lipid/100 g) and phospholipid (g/mg lipid) contents of rats fed 10 wt% hydrogenated coconut oil or 5, 10, or 20 wt% menhaden oil were not significantly different (Table 4.10). In addition, the content and distribution of heart phospholipid classes were similar in all four groups [9]. In contrast, the percentage of

TABLE 4.10 Animal Feeding Trials: Summary of Effects of Marine Oil and Eicosapentaenoic Acid Consumption on Cardiac and Aorta Characteristics.

Reference	Diets: Compared To	Cardiac Lipid Content	Cardiac Lipid Fatty Acid Composition	Aortic Lipid Fatty Acid Composition	Aortic Lipid PGI$_2$-like activity	6-Keto-PGF$_1$ synthesis in aorta
I. RATS						
(6) 1.22 wt% safflower oil (S.O.) 2.28 wt% triolein mix (TM) 1.50 wt% enriched menhaden	1.22 wt% S.O. 3.78 wt% TM			phospholipids LA AA EPA DHA NSD ↓21% ↑288% ↑347% (wt%)		(ng/ml) NSD
1.22 wt% S.O. 2.28 wt% TM 1.09 wt% enriched shark oil	1.22 wt% S.O. 3.78 wt% TM			phospholipids LA AA EPA DHA NSD ↓17% ↑155% ↑265% (wt%)		(ng/ml) NSD
(8) 6.4g/100g menhaden oil	6.9% soybean oil			LA AA EPA DHA ↓29% ↓41% ↑650% ↑211% (mole%)		
(9) 5 wt% menhaden oil	10 wt% hydrog. coconut oil (HCO)	(g lipid/100g)(total lipids) NSD	LA AA EPA DHA ↓18% ↓35% ↑58% (mole%)			
10 wt% menhaden oil	10 wt% HCO	NSD	↓7% ↓32% ↑80%			
20 wt% menhaden oil	10 wt% HCO	NSD	↓47% ↓26% ↑68% (mole%)			
(14a) 5 en% corn oil	5 en% corn oil 5 en% cod liver oil			phospholipids LA AA EPA DHA ↑58% ↓34% - - (mole%)		
(14b) 5 en% cod liver oil	5 en% corn oil			phospholipids LA AA EPA DHA NSD ↓31% ↑ - (mole%)		
(17) 12.5% saturated fat + 2.5% cod liver oil	15% saturated fat				12 min. 30 min. incub. ↓33% incub. ↓31%	
10.0% saturated fat + 5% cod liver oil	15% saturated fat				12 min. 30 min. incub. ↓42% incub. ↓35% (ng/min/cm^2) ↓	
(18) 50 en% sunflower seed oil	50 en% cod liver oil		in heart mitochondria ↓LA ↓AA ↑EPA ↑DHA (mol/100 mol)			
(19) 10 wt% cod liver oil	basal diet: 5 wt% fat					

Effects of Dietary n-3 PUFAs of Fish Oils   135

			phospholipids	phospholipids (μmol/ml) LA AA EPA DHA	(mg/ml)	(ng/piece of aorta)
(20)	45 enz% cod liver oil 5 enz% sunflower seed oil	50 enz% sunflower seed oil			↓	↓ 50%
(21)	1% sardine oil 9% hydrogenated beef tallow (HBT)	fat-free		NSD ↓22% NSD ↑300% ↑43%		
	5% sardine oil 5% HBT	fat-free		NSD ↓39% ↓14% ↑483% ↑50%		
	1% sardine oil 9% HBT	1% corn oil 9% HBT		NSD ↓45% NSD ↑233% ↑104%		
	5% sardine oil 5% HBT	5% corn oil 5% HBT		NSD ↓66% ↓25% ↑775% ↑113% (%)		
(26)	0.7% free EPA	0.7% linoleic acid				(ng/mg aorta) ↓47%
(28)	basal diet + 81 mg EPA ethylester/kg body wt	basal diet + water				(ng/mg aorta/10 min) ↑

II. RABBITS
(29) 60g/kg Max EPA   60g/kg corn oil   (ng/mg aorta/10 min) ↑

III. PIGS   (% in dry matter)
(30) 10% mackerel oil   10% olive oil   ↑

IV. CATS   phospholipids
                                                  LA AA EPA DHA
(31) 8% calories menhaden oil   basal diet   ↓ NSD NSD
                                                       (mole%)

                                          phospholipids
                                       LA AA EPA   DHA
                                       NSD NSD ↑900% ↑542%

Comparisons:

NSD: no significant difference (p ≤ 0.05) between comparison times
→ ↓ : significant decrease between comparison times
← ↑ : significant increase between comparison times

lipids in myocardial muscle dry matter increased in pigs fed a diet containing mackerel oil, compared to pigs fed olive oil [30].

Following fish oil consumption, significant changes in the fatty acid composition of cardiac lipids were observed and included increases in EPA and DHA [9,19,21] and decreases in LA [9,19,21,31]. Although decreases in AA content were observed [9,19,21], the level of this fatty acid remained essentially unchanged in other feeding trials [21,31]. As shown in Table 4.11, when rats were fed a 1% or 5% sardine oil diet, the n-3 PUFAs replaced LA rather than AA in cardiac phospholipids [21].

The contrasting effects of fish oil consumption on total cardiac lipid AA and LA contents were noted when rats fed 5, 10, and 20 wt%

TABLE 4.11 Effect of Feeding Fat-free, Corn Oil and Fish Oil Containing Diets on the Fatty Acid Composition of Rat Heart Phospholipids. Mean ± S.D.; N=3-6.

Dietary Fat		Fatty Acid Content of Heart Phospholipids (%)			
%	Source	LA	AA	EPA	DHA
	fat-free diet	13.4 ± 1.5	18.5 ± 1.9	0.30 ± 0.0	7.75 ± 0.6
1% 9%	Sardine oil HBT*	10.4 ± 0.5[a,b]	17.8 ± 0.6	1.20 ± 0.2[a,b]	13.8 ± 1.6[a,b]
5% 5%	Sardine oil HBT	8.15 ± 0.1[a,b]	16.0 ± 0.4[a,b]	1.75 ± 0.3[a,b]	19.4 ± 0.1[a,b]
1% 9%	Corn oil HBT	18.8 ± 1.3[a]	20.3 ± 2.3	0.36 ± 0.1	9.2 ± 1.2
5% 5%	Corn oil HBT	24.2 ± 0.8[a]	21.2 ± 1.0	0.20 ± 0.0[a]	9.10 ± 0.1[a]

* hydrogenated beef tallow

[a] significantly different ($p \leq 0.02$) from fat-free

[b] significantly different ($p \leq 0.02$) from the same percent of corn oil

Source: Ref. 21.

TABLE 4.12 Fatty Acid Composition of Total Lipids from Hearts of Rats Fed Incremental levels of Menhaden Oil.

Fatty Acid	Dietary Fat			
	10 wt% HCO* (N=12)	5 wt% MO+ (N=6)	10 wt% MO (N=6)	20 wt% MO (N=6)
	Fatty Acid Composition of Total Lipids (mole%)			
LA	22.17 ± 3.61	18.17 ± 2.73	18.43 ± 0.91	11.79 ± 1.39
AA	17.11 ± 2.31	11.42 ± 0.36	11.7 ± 1.08	11.03 ± 1.14
EPA	trace	2.28 ± 1.25	3.75 ± 0.64	3.47 ± 0.22
DHA	7.82 ± 2.38	12.39 ± 1.43	14.10 ± 2.24	20.99 ± 1.13

* HCO: hydrogenated coconut oil
+ MO: menhaden oil

Source: Ref. 9.

menhaden oil were compared to animals fed 10 wt% hydrogenated coconut oil. The AA level was decreased by aproximately one-third at all three levels of fish oil consumption, while the reduction in LA content varied between 18% with 5 wt% and 10 wt% fish oil consumption and 47% with 20 wt% fish oil consumption [9] (Table 4.12).

In addition, the extent of incorporation of n-3 and n-6 PUFAs differed among various cardiac lipid classes. Compared to phosphatidylcholine (PC), phosphatidylethanolamine (PE), phosphatidylserine (PS), and phosphatidylinositol (PI), the neutral glycerides showed a greater incorporation of EPA at all levels of menhaden oil intake. The greatest change in DHA levels was observed in PE. AA content decreased in PE, PC and PS/PI. The reduction of AA in PE was progressive as dietary menhaden oil increased, while the response in PC and PS/PI was similar at all fish oil intake levels. LA was the most resistant to dietary modification. Its content in PC was unchanged with less than 10 wt% menhaden oil consumption; in PE and in neutral glycerides LA decreased significantly only when 20 wt % of the diet was menhaden oil. The LA content in PS/PI did not change even when the diet containing the greatest amount of menhaden oil was consumed [9].

Consequences of Experimentally Induced Myocardial Infarctions

Coronary artery thrombosis was experimentally induced in dogs by electrically stimulating the left circumflex coronary artery for twenty-four hours. Thirty-six to forty-five days before the study, a group of the animals received 25% of their calories as menhaden oil. Dogs fed commercial chow for one to seven days before myocardial infarctions served as a control group. Among the survivors, those receiving fish oil maintained a more normal ECG pattern. The mean myocardial infarct size and the weight of the thrombi were compared in the two groups. While the mean weights of the thrombi formed were not significantly different, the dogs fed menhaden oil had a 52% smaller infarct size [32]. The evidence suggested fish oil may protect against myocardial damage occurring during infarctions.

EFFECTS OF MARINE OIL AND n-3 PUFA CONSUMPTION ON AORTIC TISSUE

Aortic Lipid Fatty Acids

Following consumption of fish oil by rats and rabbits, the EPA and DHA content of aortic phospholipids and total lipids increased by several hundred percent [6,8,14a,14b,29]. The AA content most frequently decreased by one-fifth [6] to almost one-half [18] of its initial level. The LA level remained unaltered [6,14b,29], although an increase [14a] and a decrease [8] also were reported.

Prostacyclin Production

Prostacyclin production by the aorta was determined by bioassay or radioimmunoassay. In the bioassay, the anti-aggregatory activity of the aorta was studied by incubating samples and determining the ability of the media solutions to decrease platelet aggregability in the presence of an inducing agent. Significantly less anti-aggregatory activity in the aorta was observed in rats fed fish oil [17,18,20]. However, rats which were given EPA ethylester for eight weeks showed about three times as much PGI_2-like activity than did the comparison group not provided the supplement. The mechanism of the enhanced production was not determined [28].

The concentration of 6-keto-PGF$_1\alpha$ (ng/ml), the stable metabolite of PGI$_2$, was not altered in rats fed n-3 PUFA enriched menhaden or shark liver oils [6], while 6-keto-PGF$_1\alpha$ (ng/piece of aorta) decreased 50% in rats fed cod liver oil as opposed to sunflower seed oil [20]. In another trial, PGI$_2$ production by aortic tissues was measured as the increment of 6-keto-PGF$_1\alpha$ levels in PRP after perfusion through the aorta. Rats fed 5 en% cod liver oil rather than corn oil had significantly reduced levels [14b]. In contrast, the rats that were fed EPA ethylester and had elevated PGI$_2$-like activity also had elevated 6-keto-PGF$_1\alpha$ levels in the aorta [28].

The consumption of fish oils decreased prostacyclin levels in the serum of clotted blood. The concentrations of 6-keto-PGF$_1\alpha$ were compared in rats fed varying amounts of safflower, linseed, coconut, or cod liver oil. When these oils constituted 40 en% of the diets, there were significant differences between groups; those fed cod liver oil produced 73% less 6-keto-PGF1α than the safflower oil fed group [16].

17-6-keto-PGF$_1\alpha$ is a stable metabolite of PGI$_3$. Significant amounts were not detected in rat feeding trials designed to measure its concentration in the aorta [20,28].

In brief, fish oil consumption most often decreased the formation of PGI$_2$.

Conversion of Exogenous Fatty Acids to Prostaglandins

The conversion of exogenous AA to PGI$_2$ was measured by incubating pieces of aorta with labeled AA. Rats receiving 0.7 wt% free EPA did not show a conversion rate significantly different from animals fed the same amount of linoleic acid [26]. Aortae of rats consuming 81 mg EPA ethylester/kg body weight did not convert labeled AA to PGI2 (detected as 6-keto-PGF1α) at a rate significantly different from the aortae of rats not receiving the supplement.

In these two feeding trials, the conversion of exogenous EPA to prostaglandins by aortae also was examined. No EPA-derived prostaglandins were detected in either the treatment or control groups [26,28].

Aortic Lipid Metabolism

It was hypothesized that EPA affects cholesterol esterase, and hence lipid metabolism and the formation of atheromatous lesions, by changing the membrane properties of aorta cells or by altering prostanoid metabolism [27]. The activities of aortic enzymes involved in cholesterol ester hydrolysis and synthesis were compared in rats fed a basal diet and rats fed diets supplemented with EPA and DHA. Compared to the control group, the activities of acid and neutral cholesterol esterase (enzymes involved in cholesterol ester hydrolysis) were reduced by half in those groups receiving EPA. In the DHA-supplemented group, no significant differences were measured. The activities of acyl-CoA synthetase and acyl-CoA:cholesterol acyltransferase (enzymes involved in cholesterol ester synthesis) were comparable in all three groups.

Aortic Contractions

After three weeks on diets containing either soybean or menhaden oil, rats were killed and their aortae excised. Strips of aorta were mounted in organ chambers and their isometric contractions recorded. Aortic strips from rats fed menhaden oil were significantly less responsive to the contractile effects of norepinephrine (Figure 4.4). Treatment of the aortic strips with indomethacin (a cyclooxygenase inhibitor) decreased the responsiveness of both groups. It was suggested that norepinephrine caused a release of AA in the aortae of the control group rats, while both AA and EPA were released from the aortae of rats fed fish oil. In the control group, AA was converted to dienoic prostaglandins which contributed to the contractile response induced by norepinephrine. In rats fed fish oil, the release of EPA inhibited the conversion of AA to prostaglandins and decreased aortic contractions [8].

Arterial Thrombosis Tendency

The effect of fish oil consumption on arterial thrombosis was studied in rats by measuring the obstruction time of a cannula inserted into the abdominal aorta. A longer obstruction time was considered to reflect a diminished tendency for thrombosis. Rats fed 45 en% cod liver oil and 5 en% sunflower seed oil had a mean

Effects of Dietary n-3 PUFas of Fish Oils

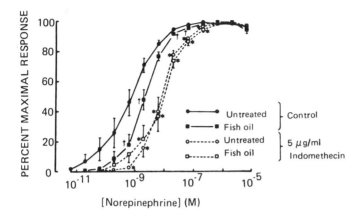

FIGURE 4.4 Dose response of norepinephrine of aortic strips from fish oil-fed and control rats. Mean ± SEM; N=6. A statistically significant decrease (p<0.05) exists between the responses in the presence and absence of indomethacin. A statistically significant decrease (p<0.05) exists between the responses in aortae from control and fish oil-fed rats. (From Ref. 8, reprinted by permission from Geron-X, Inc. Publishers.)

obstruction time (hours) 61% longer than did rats fed 5 en% sunflower seed oil. However, the mean obstruction time was not significantly different from rats fed 50 en% sunflowerseed oil [20].

EFFECTS OF MARINE OIL OR n-3 PUFA CONSUMPTION ON OTHER BLOOD CHARACTERISTICS

Hemoglobin

Pigs fed a 10% mackerel oil diet had a mean hemoglobin level 20% less than those fed olive oil [30].

Blood Pressure

The effects of a fish oil diet on blood pressure were conflicting. No significant differences in systolic blood pressure, as measured by tail plethysmography, were observed among rats consuming diets containing various levels of either safflower, linseed, coconut, or cod liver oil [16]. The blood pressure of gerbils whose mothers had consumed menhaden oil during their

TABLE 4.13 Animal Feeding Trials: Summary of Effects of Marine Oil and Eicosapentaenoic Acid Consumption on Hepatic Characteristics.

Reference	Diets	Compared To	Weight	Lipid Content	Lipid Fatty Acid Composition	TG	Chol.	Lipid Peroxide
I. RATS								
(4)	100g/kg herring oil	100g/kg maize oil			microsomes LA AA EPA DHA ↓80% ↓64% ↑ ↑580% (%)			
(5)	5% 3:7 corn-menhaden mix	20% coconut oil			microsomes LA AA EPA DHA ↑19% ↓59% ↑ ↑379% (wt%)			
(6)	1.22 wt% safflower oil 2.28 wt% triolein mix 1.50 wt% enriched menhaden oil	1.22 wt% safflower oil 3.78 wt% triolein mix			microsomes LA AA EPA DHA NSD ↓48% ↑133% ↑200% (wt%)			
	1.22 wt% safflower oil 2.70 wt% triolein mix 1.09 wt% enriched shark oil	1.22 wt% safflower oil 3.78 wt% triolein mix			microsomes LA AA EPA DHA NSD ↓34% ↑73% ↑305% (wt%)			
(13)	5% safflower oil 5% hydrogenated coconut oil	5% safflower oil 10% hydrogenated coconut oil			microsomes LA AA EPA DHA ↑34% ↓38% ↑235% ↑277% (wt%)			
	5% safflower oil 10% menhaden oil	5% safflower oil 10% hydrogenated coconut oil			microsomes LA AA EPA DHA ↑36% ↓45% ↑229% ↑222% (wt%)			
(16)	20% cod liver oil	20% safflower oil			phosphatidylethanolamine LA AA EPA DHA NSD ↓37% ↑ NSD (%)			
(19)	10% cod liver oil	basal diet: 5 wt% diet			microsomes LA AA EPA DHA → → ↑ ↑ (mol/100mol) mitochondria LA AA EPA DHA → → ↑ ↑ (mol/100mol)			

Effects of Dietary n-3 PUFAs of Fish Oils

	Diet 1	Diet 2			phospholipids	
					LA AA EPA DHA (%)	
(21)	1% sardine oil 9% hydrogenated beef tallow (HBT)	fat-free		NSD	↓28% ↓39% ↑ ↑61%	
	5% sardine oil 5% HBT	fat-free		NSD	↓45% ↓49% ↑ ↑183%	
	1% sardine oil 9% HBT	1% corn oil 9% HBT		NSD	↓63% ↓59% ↑ ↑116%	
	5% sardine oil 5% HBT	5% corn oil 5% HBT		NSD	↓80% ↓70% ↑ ↑307%	
			phospholipids ↑24% (mg/g)			
(22a)	3 wt% PUFA mix from squid liver	3 wt% methyl-oleate			NSD	NSD (μmol MDA/100g) ↑51%
			(g)(g/100g wt) tot. lipids	(mg/g)		(μmol MDA/100g)
(25b)	5% wt% PUFA mix from squid liver	5% lard	NSD ↑23% ↓30% (g/100g)	↓42%		↑260%
	2% methyl-oleate + 3% PUFA mix	5% lard	NSD ↑22%	↓19%		↑123% (μmol MDA/100g)
	4% methyl-oleate + 1% PUFA mix	5% lard	NSD ↑12%	↓10%		NSD
					phospholipids phosphatidylcholine	
					LA AA EPA DHA LA AA EPA DHA	
(23)	5% fish oil	fat-free diet			NSD NSD NSD ↑305% ↑246%	(μmol/g) ↑38%
(24)	basal diet + 15 wt% MaxEPA	basal diet	(% body wt) ↑14%			
II. RABBITS					LA AA EPA DHA	
(29)	60g/kg MaxEPA	60g/kg corn oil			↓67% ↓9% ↑160% ↑93%	
III. PIGS					LA AA EPA DHA	
(30)	10% mackerel oil	10% olive oil			NSD ↓ ↑ ↑ (wt%)	NSD
IV. CATS					LA AA EPA DHA	
(31)	8% calories as menhaden oil	basal diet			↓ ↓ ↑ ←	NSD (mole%)

Comparisons:

NSD: no significant difference ($p \leq 0.05$) between comparison times

→ ↓ : significant decrease between comparison times

↑ ← : significant increase between comparison times

pregnancy and who consumed the same diet for two months after weaning did not differ significantly from that of gerbils who did not receive the fish oil [33]. In contrast, rats fed a saturated fat diet for six weeks experienced no significant change in systolic blood pressure, while those fed saturated fat and cod liver oil experienced blood pressure increases proportional to their consumption of cod liver oil [17].

EFFECTS OF MARINE OIL OR n-3 PUFA CONSUMPTION ON HEPATIC CHARACTERISTICS

Hepatic Weight

Liver weights in relation to total body weight were increased in both rats and pigs fed marine oils. The livers of rats fed a 5% squid liver PUFA mix were 23% heavier than the livers of rats fed a 5% lard diet. As shown in Table 4.14, when part of the PUFA mix was replaced with methyl oleate the increase in liver weight was correspondingly

TABLE 4.14 Weight of the Livers of Rats Fed Lard and a Squid Liver Polyunsaturated Fatty Acid Mix. Mean ± S.E.M.; N=6.

Diet	Liver Weight (g/100g body wt)
basal diet +5% lard	6.37 ± 0.24[a]
+5% methyl-oleate	7.11 ± 0.23[a,c]
+4% methyl-oleate 1% PUFA mix	7.15 ± 0.16[c]
+2% methyl-oleate 3% PUFA mix	7.79 ± 0.21[b,c]
+5% PUFA mix	7.81 ± 0.10[b]

[a,b,c] Values in the same column not sharing a common superscript letter are significantly different ($p<0.05$)

Source: Ref. 22.

less [22b]. Pigs fed a 10% mackerel oil diet had a mean liver weight 14% greater than that of pigs fed olive oil [30].

Hepatic Lipids

The hepatic lipid content was rarely altered by marine oil consumption. Only rats fed a diet containing 5% squid liver PUFA mix showed a 30% reduction in lipid content when compared to rats fed the same proportion of lard [22b].

Hepatic Lipid Fatty Acids

The consumption of fish oil was usually accompanied by changes in the fatty acid composition of hepatic lipids. Increases, often of several hundred percent, in EPA and DHA levels were observed [4,5,6,10,13,21,23,29]. Decreases in AA content to one-third to two-thirds of the initial value were frequent [4,6,9,13,16,21]. Reports of the change in LA content differed, i.e. LA increased [5,13] and decreased [4,19] in hepatic microsomes and mitochondria. The differences in fatty acid compositions of liver microsomal lipids of rats fed varying amounts of safflower, hydrogenated coconut, and menhaden oils is shown in Table 4.15.

TABLE 4.15 Fatty Acid Composition of Liver Microsomal Lipids of Rats Fed Safflower, Hydrogenated Coconut, and Menhaden Oils. Mean ± S.D.; N=6.

Fatty Acid	Dietary Fat		
	5% SAF* + 10% HCO	5% SAF + 5% HCO + 5% ME	5% SAF + 10% ME
	Fatty Acid Composition of Liver Microsomal Lipids (wt%)		
LA	13.2 ± 1.3	17.7 ± 1.3[a]	17.9 ± 1.6[a]
AA	27.9 ± 2.1	17.2 ± 1.8[a]	15.3 ± 1.8[a]
EPA	1.4 ± 0.3	4.7 ± 1.1[a]	4.6 ± 1.0[a]
DHA	1.8 ± 0.5	6.8 ± 0.9[a]	5.8 ± 1.2[a]

* SAF, safflower oil; HCO, hydrogenated coconut oil; ME, menhaden oil.

[a] significantly different ($p < 0.05$) to content of same fatty acid in the group fed 5% SAF and 10% HCO.

Source: Ref. 13.

Measurable changes in fatty acid composition of liver lipids were observed soon after the initiation of marine oil consumption. After one and two days of feeding 100 g/kg herring oil to rats, the microsomal fatty acid composition changed, with marked increases in the proportion of EPA and DHA. However, for a more constant change in composition, ten days of feeding were required [4]. When rats fed a fat-free diet for fourteen days were provided with a 5% fish oil diet for the same period, the level of 20:3n-6 in hepatic phospholipids quickly decreased while the proportion of n-3 PUFAs increased. Plateau levels of n-3 PUFAs were observed by the fourth day of fish oil feeding. By the sixth day of fish oil consumption, 20:3n-6 almost disappeared. When the dietary process was reversed, the level of n-3 PUFAs dropped at a rate lower than their rate of increase. On the sixth day of a fat-free diet, the n-3 PUFA levels fell to half the peak value [23].

In a feeding trial in which rats were fed either a fat-free diet or a diet containing fish or corn oil and hydrogenated beef tallow, the total proportion of unsaturated fatty acids in the liver was 53% to 60%, irrespective of the dietary fat sources. To reach these levels, endogenous unsaturated fatty acids (16:1, 18:1, 20:3n-9) in rats fed fat-free or hydrogenated fat diets increased. Regardless of the amounts and types of dietary unsaturated fatty acids, the total amount of fatty acids with more than three double bonds remained at a constant level of 31% to 35% in the liver. If rats were fed diets which did not contain polyunsaturated fatty acids, the levels of endogenous 20:3n-9 increased. Thus, hepatic fatty acid levels appear to be physiologically important, and the desaturation of fatty acids is affected by dietary fatty acids [21].

Briefly, the hepatic lipid content was not easily altered by a marine oil diet. However, changes in the fatty acid composition of hepatic lipids occured readily; increases in the proportions of EPA and DHA and decreases in AA content were observed frequently while the change in LA content was more variable. These alterations in fatty acid composition and their reversal were measureable within days of a dietary modification.

Hepatic Triglycerides

The effects of marine oil consumption on the level of hepatic triglycerides remains unclear. Hepatic triglyceride levels were not altered when rats were fed a squid liver PUFA mix rather than methyl oleate [22a] and when pigs were fed mackerel oil rather than olive oil [30]. On the other hand, rats fed a basal diet supplemented with 15 wt% MaxEPA had a 38% greater triglyceride concentration (mol/g) than did rats whose diets were not supplemented [24].

Cholesterol

Decreases in hepatic cholesterol (mg/g) were proportional to marine oil intake. A 42% drop was observed among rats fed a squid liver PUFA mix compared to rats fed the same proportion of lard. With increasing substitution with methyl oleate, smaller mean decreases in cholesterol were observed [22b].

Enzyme Activity

A major mechanism by which n-3 PUFAs may depress tissue AA is via inhibition of the enzymes converting dietary LA to AA. this applies particularly to the $\Delta 6$ desaturase - the rate-limiting step in this pathway. Significantly, fish oil consumption lowered the activities of $\Delta 6$-desaturase, elongase, and $\Delta 5$-desaturase, all enzymes required for the synthesis of AA from LA. When compared to rats fed a diet containing 20% coconut oil, the activity of $\Delta 6$-desaturase (nmol enzyme product/min/mg protein) in rats fed a diet containing 5 wt% 3:7 corn oil:menhaden oil mix was reduced by a mean of 43%; the activity of elongase was reduced by 52% and that of 5-desaturase by 64% (Table 4.16). It was suggested that the higher levels of EPA and DHA in the fish oil diet inhibited the enzymes [5]. In another feeding trial, a mean 29% reduction in the activity of 6-desaturase also was observed in rats fed menhaden rather than safflower and coconut oil for thirty-three weeks [13]. These reductions in activity depressed the conversion of LA to AA and probably accounted for the increases in LA and decreases in AA [5,13]. (Table 4.17.)

When hydrogenated coconut oil was replaced totally or in part by menhaden oil in the diet of rats, the activity of 9-desaturase, which

TABLE 4.16 Effect of Dietary Fat on Liver Microsomal Enzyme Activities. Mean ± S.D.; N=6.

Enzyme	Dietary Fat		
	20% CO[*]	5% CME	20% CME
	Enzyme Activity (nmol/min/mg protein)		
6-desaturase	0.284 ± 0.048[a]	0.162 ± 0.014	0.182 ± 0.040
elongase	0.178 ± 0.036	0.086 ± 0.012	0.094 ± 0.018
5-desaturase	0.056 ± 0.010	0.020 ± 0.008	0.022 ± 0.008

[*] CO, coconut oil; CME, 3:7 corn-menhaden oil mixture.

[a] enzyme activities with 20% CO are significantly different (p<0.01) to those measured with 5% and 20% CME. The values for 5% and 20% CME are not significantly different.

Source: Ref. 5.

TABLE 4.17 Effect of Dietary Fat on the Activity of 6-desaturase of Rats. Mean ± S.D.; N=6.

Dietary Fat	6-desaturase activity in hepatic microsomes (nmol 18:3n6/min/mg microsomal protein)
5% SAF[*] + 10% HCO	0.36 ± 0.04[a]
5% SAF + 5% HCO + 5% ME	0.27 ± 0.03
5% SAF + 10% ME	0.20 ± 0.03

[*] SAF, safflower oil; HCO, hydrogenated coconut oil; ME, menhaden oil.

[a] all values in column are statistically different (p<0.05) from each other.

Source: Ref. 13.

Effects of Dietary n-3 PUFAs of Fish Oils 149

converts 18:0 to 18:1, decreased (Table 4.16). Total replacement of the vegetable oil by the marine oil resulted in a 44% decrease in activity, while a 25% drop was evident when half of it was replaced. Since the fish oil diet contained greater levels of 16:1 and 18:1, it was suggested that the activity of 9-desaturase may be controlled by the dietary intake of these fatty acids or of highly polyunsaturated fatty acids [13].

No significant differences in the amount or activity of carnitine palmitoyltransferase in the absence of malonyl CoA were observed in rats fed MaxEPA and safflower oil. However, marked differences were found in the sensitivity of the enzyme to inhibition by malonyl CoA. The concentration of malonyl CoA required to suppress the enzyme activity by 50% was 139% greater in rats fed MaxEPA than among rats fed safflower oil, signifying a rise in mitochondrial oxidation [24].

In the same rats, the activity of glycerophosphate acyltransferase was also assayed. Total activity measured with palmitoyl-CoA and oleyl-CoA increased by 59% and 167% in the MaxEPA-fed rats compared to those fed a basal diet.

The activity of 5' nucleotidase (nmoles Pi liberated/mg protein/min) in liver plasma membranes was doubled in rats fed 10 wt% menhaden oil compared to animals fed hydrogenated coconut oil. This increased activity could have been a result of changes in membrane fluidity brought about by the incorporation of dietary n-3 PUFAs into membranes or the action of the fatty acids on the conformation of enzyme [10].

The effect of fish oil consumption on hepatic oxidative metabolism was studied by feeding rats diets containing 100 g of herring oil/kg diet and measuring the rate of oxidative demethylation of aminopyrine in the hepatic endoplasmic reticulum. The rate was 30% greater among rats fed herring oil than among those fed the same amount of maize oil. When rats consumed herring oil in which the n-3 PUFAs had been destroyed by irradiation, the rate of oxidative demethylation did not increase. This suggests that the incorporation of n-3 PUFAs into hepatic microsomes, which occured when the

TABLE 4.18 Liver Lipid Peroxide in Rats Fed Lard and a Squid Liver Polyunsaturated Fatty Acid Mix. Mean ± S.E.M.; N=6.

Diet	Liver Lipid Peroxide (μmol MDA/100g liver)
basal diet +5% lard	33.6 ± 2.8[a]
+4% methyl-oleate 1% PUFA mix	43.7 ± 6.8[a]
+2% methyl-oleate 3% PUFA mix	74.8 ± 17.0[b]
+5% PUFA mix	120.7 ± 22.8[b]

[a,b] values in same column not sharing a common superscript are significantly different ($p<0.05$)

Source: Ref. 22.

nonirradiated oil was fed, could influence hepatic oxidative metabolism [4].

Lipid Peroxide

Lipid peroxide (measured as umol MDA/100 g) was increased 51% in rats fed a squid liver PUFA mix rather than methyl oleate [22a]. Increases in lipid peroxide proportional to the amount of PUFA mix consumed were observed when rats fed the PUFA mix alone or combined with varying proportions of methyl oleate were compared to those fed lard (Table 4.18). The increases of serum peroxide were not due to the ingestion of peroxide-rich oil but rather to the de novo formation of peroxide. This response may signify an increased requirement for dietary antioxidant [22b].

EFFECTS OF MARINE OIL OR n-3 PUFA CONSUMPTION ON LUNG LIPIDS

Lung Lipid Content

Fish oil consumption did not alter the phospholipid content of lung microsomes of rats or guinea pigs fed diets containing 10%

hydrogenated coconut oil or half hydrogenated coconut oil and half menhaden oil [7].

Lung Lipid Fatty Acids

Significant changes were observed in the fatty acid composition of lung microsomal phospholipids. The EPA levels (nmoles/mg protein) in rats that consumed the diet containing menhaden oil increased more than 200% and DHA levels increased more than 154%. AA dropped by 69% and LA remained unchanged [7b]. Guinea pigs consuming menhaden oil had elevated levels of EPA and DHA, but LA and AA content was not changed [7]. In both species, the fatty acid composition of microsomal neutral lipids was not affected by the dietary treatments [7]. (Table 4.19.)

Diets containing either n-3 PUFA enriched menhaden oil (abundant in EPA) or n-3 PUFA enriched shark liver oil (abundant in DHA) for three weeks caused similar changes in the fatty acid composition of

TABLE 4.19 Fatty Acid Composition of Lung Microsomal Phospholipids from Rat and Guinea Pig After Receiving a Diet Enriched in Menhaden Oil for 3 Weeks. Mean ± S.D.

Fatty Acid	Rat		Guinea Pig	
	Control N=5	Menhaden Oil Enriched Diet N=6	Control N=6	Menhaden Oil Enriched Diet N=7
	Fatty Acid Composition of Lung Microsomal Phospholipids (nmoles/mg protein)			
LA	31.3 ± 3.7	29.1 ± 4.7	188.3 ± 16.4	190.4 ± 6.0
AA	99.3 ± 7.3	30.8 ± 6.5^a	131.9 ± 20.7	114.0 ± 12.7
EPA	9.1 ± 0.6	27.4 ± 4.0^a	ND*	82.8 ± 7.7
DHA	8.4 ± 2.5	21.3 ± 2.3^a	ND	32.5 ± 3.4

* ND, not detected

^{a} significantly different ($p<0.001$) to fatty acid content in control group.

Source: Ref. 7.

rat lung microsome phospholipids. Compared to a control group, EPA and DHA levels increased, the latter by more than 300%, and the AA content fell by approximately 30%. The LA content did not differ significantly between the three groups [6].

Cholesterol

The cholesterol content of lung microsomes of rats and guinea pigs fed either hydrogenated coconut oil or hydrogenated coconut and menhaden oils was not altered by fish oil consumption [7a,7b].

Prostaglandin Formation

As shown in Figure 4.5, lung microsomes from rats fed either 1.5 wt % n-3 PUFA enriched menhaden oil (abundant in EPA) or 1.09 wt % n-3 PUFA enriched shark liver oil (abundant in DHA) converted exogenous ^{A}A to PGE, TXB^2, and PGF$_2$ (pmoles/mg protein) in amounts similar to the control group [6]. When rats and guinea pigs were fed diets that contained either 10 wt% hydrogenated coconut oil or 5 wt%

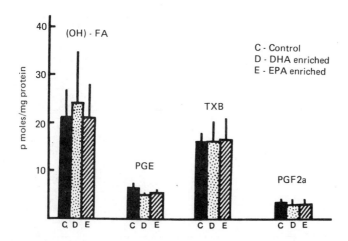

FIGURE 4.5 Conversion of exogenous arachidonic acid to eicosanoids by lung microsomal fractions from rats fed EPA and DHA. C: control (n=10); D: received n-3 PUFA enriched shark liver oil (n=10); E: received n-3 PUFA enriched menhaden oil (n=10). (From Ref. 6, reprinted by permission from Pergamon Press, Inc.)

Effects of Dietary n-3 PUFAs of Fish Oils 153

hydrogenated coconut oil and 5 wt% menhaden oil, the lung microsomes of the animals fed menhaden oil showed 35% to 50% less capacity to convert ^AA to PGE, thromboxane, or PGF_2 [7]. Thus only the higher levels of fish oil consumption altered prostaglandin synthesis by lung microsomes.

EFFECTS OF MARINE OIL OR n-3 PUFA CONSUMPTION ON KIDNEY LIPIDS

Lipids

The fatty acid composition of kidney lipids was modified by fish oil consumption (Table 4.20). In the kidney medulla, the concentrations of both LA and AA decreased by more than one-third and the EPA level increased in rats fed menhaden rather than corn oil. Similar changes were reported in the kidney cortex [11]. Kidney LA and AA decreased by more than one-third in rats fed cod liver oil compared to those consuming safflower oil. EPA and DHA were not detected in the vegetable oil [16]. In contrast, AA in kidney phospholipids decreased by one-third in the rats fed fish oil but no significant difference in LA was observed between the two groups. This decrease in AA was accompanied by a significant incorporation of EPA and DHA in the kidneys of rats fed cod liver oil [16].

Prostaglandin Production

Gas chromatography and mass spectrometry were used to determine the in vitro production (ng/mg tissue) of $PGF_2\alpha$ by homogenates of kidney medullae and cortices of rats fed menhaden or corn oil. As shown in Table 4.21, $PGF_2\alpha$ levels were reduced by approximately one-half in the medullae of rats fed the fish oil. $PGF_2\alpha$ levels in the cortices of these rats were reduced by about 90% compared to those fed corn oil [11].

Radioimmunoassay was used to determine the production of 6-keto-$PGF_1\alpha$ by kidney tissue of rats fed cod liver oil or safflower oil. When both oils were consumed at the 40 en% level, cod liver oil caused a significant decrease in the synthesis of 6-keto-$PGF_1\alpha$ [16].

The in vitro production of PGE_2 by homogenates of kidney medullae and cortices was decreased by 50% in rats fed menhaden

TABLE 4.20 Concentration of n-6 and n-3 Fatty Acids in Kidney Medullae and Cortices from Rats Fed Corn Oil and Menhaden Oil Diets. Mean ± S.E.M.

Fatty Acid	Diet Treatment	
	Corn Oil[+]	Menhaden Oil
	Fatty Acid Content (µg/mg tissue)	
	Medullae	
LA	0.88 ± 0.04	0.55 ± 0.02[a]
AA	2.66 ± 0.14	1.50 ± 0.08[a]
EPA	ND[*]	0.43 ± 0.03
DH	Trace	0.43 ± 0.05
	Cortices	
LA	5.37 ± 0.51	3.50 ± 0.21[a]
AA	12.04 ± 0.84	5.12 ± 0.25[a]
EPA	ND	2.40 ± 0.23
DHA	Trace	1.53 ± 0.11

[+] Corn Oil, 5% corn oil diet
Menhaden Oil, 4% menhaden oil + 1% corn oil diets

[*] ND, none detected
Trace, less than 1% of total fatty acids

[a] concentrations of polyunsaturated fats are significantly different between diets ($p<0.001$)

Source: Ref. 11.

instead of corn oil [11] (Table 4.21). When cod liver and safflower oils were fed at the 40 en% level, rats receiving the former oil had a significantly decreased level of PGE_2 [16].

PGE_3 was identified by gas chromatography-mass spectroscopy in the homogenates of kidney medullae of rats fed menhaden oil for twenty-two weeks [12].

TABLE 4.21 Levels of PGE_2 and PGF_2 Produced by Kidney Medullae and Cortices from Rats Fed Corn Oil and Menhaden Oil Diets for 22 Weeks. Mean ± S.E.M.; N=10.

Strain[*]	Corn Oil[+]	Menhaden Oil	Corn Oil	Menhaden Oil
	PGE_2 (ng/mg tissue)		$PGF_{2\alpha}$ (ng/mg tissue)	
		Medullae		
SHR	18.8 ± 0.6[a]	9.7 ± 0.4	10.8 ± 1.0	5.7 ± 0.6
SHR/SP	17.9 ± 0.7	9.3 ± 0.6	10.3 ± 0.7	5.3 ± 0.3
WKY	18.1 ± 0.7	8.3 ± 0.5	10.6 ± 0.5	4.9 ± 0.3
		Cortices		
SHR	0.25 ± 0.05	0.12 ± 0.02	2.89 ± 0.98	0.34 ± 0.09
SHR/SP	0.25 ± 0.03	0.07 ± 0.02	1.44 ± 0.38	0.19 ± 0.07
WKY	0.28 ± 0.05	0.12 ± 0.03	4.10 ± 1.22	0.29 ± 0.09

[*] SHR, spontaneously hypertensive rat; SHP/SP, stroke-prone spontaneously hypertensive rat; WKY, Kyoto Wistar rat (normotensive)

[+] Corn Oil, 5% corn oil diet; Menhaden Oil, 4% menhaden oil + 1% corn oil diet.

[a] significant difference ($p<0.001$) by diet for all strains.

Source: Ref. 11.

In summary, the synthesis of diene prostaglandins by the kidney decreased with marine oil consumption. PGI3 formation by kidney homogenates was tentatively identified.

EFFECTS OF MARINE OIL AND n-3 PUFA CONSUMPTION ON ADIPOSE TISSUE

Adipose Lipid Fatty Acids

The fatty acid composition of total adipose tissue lipids was modified by fish oil consumption. In rats fed a 60 g MaxEPA/kg diet rather than corn oil, the LA content was 37% less, while EPA and DHA levels were significantly greater [29] (Table 4.22).

"Yellow Fat" Disease

Pigs fed a diet containing 10% mackerel oil for four weeks showed signs of moderate vitamin E deficiency. All groups showed

TABLE 4.22 Fatty Acid Composition of Adipose Tissue of Rabbits Fed Corn Oil and MaxEPA for 60 Days. Mean; N=6.

Fatty Acid	Dietary Treatment	
	Corn Oil[+]	MaxEPA
	Fatty Acid Composition (%)	
LA	38.7[a]	24.4[b]
AA	-	-
EPA	-	2.7
DHA	-	3.3

[+] corn oil, 60g corn oil/kg feed
MaxEPA, 60g MaxEPA/kg feed

[a,b] mean values without a common superscript differ significantly ($p<0.05$)

Source: Ref. 29.

degenerative changes and inflammation associated with the accumulation of lipofuscin pigment in fat depots. One animal had a yellowish discoloration of the adipose tissues. Additional disease manifestations were the diffuse degeneration of liver cells in one pig and the accumulation of lipofuscin in the liver Kuppfer cells of four animals [30].

EFFECTS OF MARINE OIL AND n-3 PUFA CONSUMPTION ON CEREBRAL AND NEUROLOGICAL CHARACTERISTICS

Cerebral Lipid Fatty Acids

The fatty acid composition of the brain lipids of cats [31] and gerbils [33] was altered by fish oil intake. Compared to controls, cats that consumed 8% of calories as menhaden oil for eighteen to twenty-four days showed a doubling of the content (mole %) of both EPA and DHA. LA and AA levels were similar in the two groups [31]. Gerbils whose mothers had consumed a commercial chow or 25% of calories as menhaden oil during pregnancy and which consumed the

TABLE 4.23 Fatty Acid Composition of Gerbil Brains after Control and EPA-rich Diets. Mean ± S.D.

Fatty Acid	Dietary Group	
	Control *	EPA
	Fatty Acid Content (%)	
LA	1.32 ± 0.01	0.76 ± 0.09[a]
AA	9.87 ± 0.17	9.10 ± 0.10[a]
EPA	0.00 ± 0.00	0.41 ± 0.007[a]
DHA	14.86 ± 1.23	17.53 ± 0.13[a]

* Control Diet: Purina rat chow to gerbil mother during pregnancy and to gerbil for 2 months after weaning.
EPA diet: Purina rat chow and 25% of calories as menhaden oil to gerbil mother during pregnancy and to gerbil for 2 months after weaning.

[a] Significantly different ($p<0.02$) from fatty acid content of control group.

Source: Ref. 33.

maternal diet for two months after weaning had significantly different brain fatty acid compositions. Animals that consumed menhaden oil had 42% less LA (as % total fatty acids), 8% less AA, and increased EPA and DHA levels [33] (Table 4.23). Decreases in LA and AA and increases in EPA and DHA were also observed in the lipids of brain microsomes and mitochondria of rats fed a diet containing 10% cod liver oil compared to those fed the control diet [19].

Consequences of Experimentally Induced Cerebral Ischemia

After experimental ischemia (by occlusion of the right and left common carotid arteries) and reperfusion, gerbils fed menhaden oil as a source of 25% of calories and those fed a commercial chow did not differ significantly in the synthesis of brain $PGF_2\alpha$, PGE_2, 6-keto-$PGF1\alpha$, and thromboxane, as determined by radioimmunoassay [33]. A significant accumulation of water was evident after ischemia and reperfusion in the brains of gerbils fed a control diet. No significant change in brain specific gravity was noted before and after ischemia among animals fed the fish oil [33].

Measurements of cerebral blood flow before and after experimental ischemia and reperfusion were also made in the control and menhaden oil fed gerbils. Using a hydrogen clearance technique, it was observed that only 20% of the initial cerebral blood flow (ml/100g/min) was restored in the control gerbils thirty minutes after reperfusion. The volume was not changed significantly at sixty minutes of reperfusion. In contrast, the gerbils fed fish oil had cerebral blood flow volumes similar to their pre-ischemia values after ischemia [33] (Figure 4.6).

Middle cerebral artery occlusion was experimentally performed. Adult cats were fed either a commercial diet or menhaden oil as 8% of calories for eighteen to twenty-four days. The location and extent of brain lesions caused by the occlusion were determined by the histological study of brain sections. The mean volume of brain

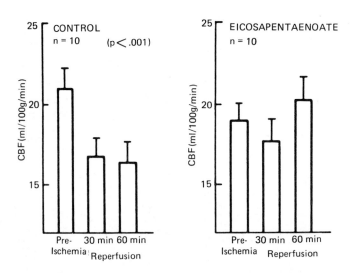

FIGURE 4.6 Cerebral blood flow (CBF) in control and eicosapentaenoic acid fed gerbils prior to ischemia and after ischemia and reperfusion. Mean ± SE. (From Ref. 33, reprinted by permission from the American Heart Association, Inc.)

necrosis was 37% less in the group treated with menhaden oil compared to the control group [31].

Two measures of neurological competence were made among the control and menhaden oil fed cats three days after temporary occlusion of the middle cerebral artery. Gait disturbances were graded on a scale ranging from no observable neurological deficit to hemiplegia. Animals treated with menhaden oil had a significantly better mean gait score. The cats' righting reflex was also graded on a scale ranging from a normal reflex to severe impairment. Those treated with menhaden oil scored significantly better [31].

These studies indicate that fish oil consumption may prevent or lessen the deleterious consequences of cerebral ischemia, including edema, diminished blood flow, brain necrosis, and impaired neurological competence.

DISCUSSION

It is apparent that fish oil consumption can cause widespread metabolic changes. In all the physiological and biochemical systems examined, alterations accompanying fish oil consumption were observed.

Much has been learned from animal feeding trials. Marine oil consumption is accompanied by changes in the fatty acid composition of lipids throughout the body. Increases in the proportion of EPA and DHA are almost universally reported, though the changes in AA and LA content are more variable. EPA uptake increases when AA and LA are reduced or eliminated from the diet. Data from platelet studies suggest that DHA is incorporated less readily than EPA into lipids.

Plasma triglycerides and cholesterol decrease with marine oil consumption. Experimental results suggest that diminished lipogenesis and increased ketogenesis are among the causes of the drop in triglyceride content. The plasma cholesterol decrease may be proportional to the dietary n-3 PUFA content. The evidence indicates that HDL cholesterol levels may be more sensitive to DHA than to EPA intake.

Fish oil consumption is also accompanied by an apparent decrease in the size of AA pools available for prostaglandin synthesis. The decreased production of MDA, HETE, and HHT by platelets, the diminished TXB_2 production during platelet aggregation, and the reduced TXB2 levels in the serum of clotted blood are evidence of this decrease. In addition, less PGI_2 is produced by vascular walls, and less 6-keto-$PGF_1\alpha$, $PGF_2\alpha$, PGE_2, and thromboxane are synthesized by lungs and kidneys.

Liver enzymes are altered by fish oil consumption. Decreased activity of the desaturases involved in the synthesis of AA from LA are among the observed changes and may be partially responsible for the above findings. Finally, fish oil consumption lessens the damage caused to cardiac tissue in infarctions and to cerebral tissue in ischemia.

In summary, many of the findings support the putative beneficial role of marine oils in preventing or lessening the risk of atherosclerosis and thrombosis.

Research is needed to determine the requirement for dietary vitamin E during increased consumption of n-3 PUFAs. Evidence that a deficiency may develop exists. Serum and lipid peroxide values increase in animals ingesting fish oils, and deleterious peroxidative changes in liver and adipose tissue cells have also been noted.

A small number of studies have monitored the time course of changes induced by dietary fish oils. Data obtained by repeating measurements or by sacrificing a fraction of the animals at each of several intervals during the course of a feeding trial are needed.

Animals on diets deficient in essential fatty acids occasionally have been used as controls. Even after a short period on such diets, abnormal metabolic alterations become apparent, limiting the value of any comparisons. Instead, animals fed appropriate amounts of vegetable oil or animal fat should be used as controls.

The research efforts suggested above could lead to greater knowledge of the mechanisms by which n-3 PUFAs act to ultimately decrease the risk of cardiovascular disease. However, cumulating evidence reveals differences in the response of tissues and species to dietary n-3 PUFAs. Differences between dietary habits of animals and metabolism in tissues suggest caution in extrapolating dietary results obtained from short-term animal trials to human behavior.

Animal feeding trials generally support the observations made with humans. However, some biochemical differences exist that make extrapolations of observations from animal studies to humans tenuous and uncertain. Animals apparently convert EPA to PGI_3 much less effectively than do humans, perhaps reflecting a lower peroxide tone. In addition, platelets from various species, including humans, behave differently in response to agonists [34]. The rates and extents of incorporation of EPA and DHA into animal tissues show species and organ variability [9].

Nevertheless, animal feeding trials are invaluable in assessing specific and detailed biochemical effects of n-3 PUFAs. Dietary intake can be carefully controlled and changes in the fatty acid pools and eicosanoid metabolism and enzyme activities of all organs can be monitored in detail. The data thus obtained is useful in defining more precise and relevant trials with humans.

REFERENCES

1. Ackman, R.G. (1980). Fish lipids. Part I. In <u>Advances in Fish Science and Technology,</u> pp. 86-103. J.J. Connell, ed. Fishing News Books, Ltd. Farmham, Surrey, England.

2. Dyerberg, J. (1981). Platelet-vessel wall interactions: influence of diet. <u>Philos. Trans. R. Soc. Lond. B</u> 294:373-381.

3. Dyerberg, J. and H.O. Bang (1982). A hypothesis on the development of acute myocardial infarction in Greenlanders. <u>Scand. J. Clin. Lab. Invest.</u> 42(suppl. 161): 9-13.

4. Hammer, C.T. and E.D. Wills. (1979). The effect of dietary fats on the composition of the liver endoplasmic reticulum and oxidative drug metabolism. <u>Br. J. Nutr.</u> 41:465-475.

5. Kurata, N. and O.S. Privett. (1980). Effect of dietary fatty acid composition on the biosynthesis of unsaturated fatty acids in rat liver microsomes. <u>Lipids</u> 15(7):512-518.

6. Bruckner, G.G., B. Lokesh, B. German, and J.E. Kinsella. (1984). Biosynthesis of prostanoids, tissue fatty acid composition and thrombotic parameters in rats fed diets enriched with docosahexaenoic or eicosapentaenoic acid. Thromb. Res. 34:479-497.

7. Lokesh, B.R., G. Bruckner, and J.E. Kinsella. (1984_). Reduction in thromboxane formation by n-3 fatty acids enriched lung microsomes from rat and guinea pig following the ingestion of dietary menhaden oil. Prostaglandins Leukotrienes Med. 15:337-348.

8. Lockette, W.E., R.C. Webb, B.R. Culp, and B. Pitt. (1982). Vascular reactivity and high dietary eicosapentaenoic acis. Prostaglandins 24(5):631-639.

9. Swanson, J.E. and J.E. Kinsella. (1986). Dietary n-3 PUFA: modification of rat cardiac lipid and fatty acid composition. J. Nutr. 116:514-23.

10. Flier, J., B. Lokesh, and J.E. Kinsella. (1985). Increased 5'nucleotidase activity in plasma membranes from rat liver following ingestion of fish oil. Nutr. Res. (in press).

11. Schoene, N.W., A. Ferretti, and D. Fiore. (1981). Production of prostaglandins in homogenates of kidney medullae and cortices of spontaneously hypertensive rats fed menhaden oil. Lipids 16(11):866-869.

12. Ferretti, A., N.W. Schoene, and V.P. Flanagan. (1981). Identification and quantification of prostaglandin E_3 in renal medullary tissue of three strains of rats fed fish oil. Lipids 16(11):800-804.

13. De Scrijver, R. and O.S. Privett. (1982). Effects of dietary long-chain fatty acids on the biosynthesis of unsaturated fatty acids in the rat. J. Nutr. 112:619-626.

14(a,b). Socini, A., C. Galli, C. Colombo, and E. Tremoli. (1983). Fish oil administration as a supplement to a corn oil containing diet affects arterial prostacyclin production more than platelet thromboxane formation in the rat. Prostaglandins 25(5):693-709.

15. Socini, A., E. Tremoli, C. Colombo, C. Galli. (1983). Fish oil administration lowers prostacyclin production and the antiaggregatory activity of aortic walls more effectively than platelet aggregation and thromboxane in the rat. Adv. Prostaglandin Thromboxane and Leukotriene Res. 12:185-191.

16. Croft, K.D., L.J. Beilin, R. Vandongen, and E. Mathews. (1984). Dietary modification of fatty acid and prostaglandin synthesis in the rat. Biochim. Biophys. Acta 795:196-207.

17. Scherag, R., H.J. Kramer, and R. Dusing. (1982). Dietary administration of eicosapentaenoic and linoleic acid increases arterial blood pressure and suppresses vascular prostacyclin synthesis in the rat. Prostaglandins 23(3):369-382.

18. Ten Hoor, F., E.A.M. De Dechere, E. Haddeman, G. Hornstra, and J.F.A. Quadt. (1980). Dietary manipulation of prostaglandin and thromboxane synthesis in heart, aorta, and blood platelets of the rat. Adv. Prostaglandin and Thromboxane Res. 8:1771-1781.

19. Tahin, Q.S., M. Blum, and E. Carafoli. (1981). The fatty acid composition of subcellular membranes in rat liver, heart and brain: diet-induced modifications. Europ. J. Biochem. 121:5-13.

20. Hornstra, G., E. Christ-Hazelhof, E. Haddeman, F. Ten Hoor, D.H. Nugteren. (1981). Fish oil feeding lowers thromboxane and prostacyclin production by rat platelet and aorta and does not result in the formation of prostaglandin I_3. Prostaglandins 21(5):727-738.

21. Iritani, N., and S. Fujikawa. (1982). Competitive incorporation of dietary w-3 and w-6 polyunsaturated fatty acids into the tissue phospholipids in rats. J. Nutr. Sci. Vitaminol. 28:621-629.

22(a,b).Kobatake, Y., F. Hirahara, S. Innami, and E. Nishide. (1983). Dietary effect of w-3 type polyunsaturated fatty acids on serum and liver lipid levels in rats. J. Nutr. Sci. Vitaminol. 29:11-21.

23. Iritani, N., and R. Norita. (1984). Changes of arachidonic acid and n-3 polyunsaturated fatty acids of phospholipid classes in liver, plasma, and platelets during dietary fat manipulation. Biochim. Biophys. Acta 793:441-447.

24. Wong, S.H., P.J. Nestel, R.P. Trimble, G.B. Storer, R.J. Illman, D.L. Topping. (1984). The adaptive effects of dietary fish and safflower oil on lipid and lipoprotein metabolism in perfused rat liver. Biochem. Biophys. Acta 792:103-109.

25. Morita, I., N. Komeshima, R. Takahashi, H. Orimo, and S. Murota. (1984). A proposed method for exploring anti-aggregatory effects of eicosapentaenoic acid in the rat. Prostaglandins Leukotrienes Med. 14:123-129.

26. Morita, I., Y. Saito, W.C. Chang, and S. Murota. (1983). Effects of purified eicosapentaenoic acid on arachidonic acid metabolism in cultured murine aortic smooth muscle cells, vessel walls and platelets. Lipids 18(1):42-49.

27. Morisaki, N., M. Shinomiya, N. Matsuoka, Y. Saito, and A. Kumagai. (1983). In vivo effects of cis-5,8,11,14,17-20:5(n-3) and cis-4,7,10,13,16,19-22:6(n-3) on serum lipoproteins, platelet aggregation, and lipid metabolism in the aorta of rats. Tohuku J. Exp. Med. 141:397-405.

28. Hamazaki, T., A. Hirai, T. Terano, J. Sajiki, S. Kondo, T. Fujita, Y. Tamura, and A. Kumagai. (1982). Effects of orally administered ethylester of eicosapentaenoic acid on PGI_2-like substance production by rat aorta. Prostaglandins 23(4):557-567.

29. Vas Dias, F.W., M.J. Gibney, and T.G. Taylor. (1982). The effect of polyunsaturated fatty acids of the n-3 and n-6 series on platelet aggregation and platelet and aortic fatty acid composition in rabbits. Atherosclerosis 43:245-257.

30. Ruiter, A., W. Jongbloed, C.M. van Gent, L.H.J.C. Danse, and S.H.M. Metz. (1978). The influence of dietary mackerel oil on the condition of organs and on blood lipid composition in the young growing pig. Amer. J. Clin. Nutr. 31:2159-2166.

31. Black, K.L., B. Culp, D. Madison, O.S. Randall, and W.E.M. Lands. (1979). The protective effects of dietary fish oil on focal cerebral infarction. Prostaglandins Med. 3:257-268.

32. Culp, B.R., W.E.M. Lands, B.R. Luchesi, B. Pitt, and J. Romson. (1980). The effect of dietary supplementation of fish oil on experimental myocardial infarction. Prostaglandins 20(6):1021-1031.

33. Black, K.L., J.T. Hoff, N.S. Radin, and G.D. Deshmukh. (1984). Eicosapentaenoic acid: effect on brain prostaglandins, cerebral blood flow and edema in ischemic gerbils. Stroke 15(1):65-69.

34. Lokesh, R.B., G. Bruckner, and J.E. Kinsella. (1984). Reduction in thromboxane formation by n-3 fatty acids enriched lung microsomes from rat and guinea pig following the ingestion of dietary menhaden oil. Prostaglandins Leukotrienes Med. 15:337-48.

5
Dietary Polyunsaturated Fatty Acids and Cancer

INTRODUCTION

Cancer risk is generally higher among people who consume diets high in fat and low in fiber, vegetables, and micronutrients [1,2]. Recent studies have significantly demonstrated that the amount of dietary fat as well as the type and composition of fat influence the development of cancer [3,4,5].

There is increasing evidence that fats containing polyunsaturated fatty acids of the n-6 family (n-6 PUFAs) are conducive to the growth of tumor cells [5,6]. In experimental animals, a relationship between the consumption of n-6 PUFAs of vegetable oils and the incidence and growth of mammary tumors has been well established [7,8]. n-6 PUFAs are converted to PGE2, which may facilitate tumor growth and metastasis [9]; high tissue levels of n-6 fatty acids may promote this reaction.

To reduce the risk of cancer, it is not feasible to eliminate n-6 PUFAs completely from the diet because they are needed for normal biochemical functions and health maintenance. Furthermore, increased consumption of vegetable oils has been advocated to improve serum lipid levels and reduce coronary disease. Ideally, dietary PUFAs should exert beneficial effects on coronary disease while suppressing the development of cancer. Appropriate dietary levels of the n-3 PUFAs of fish oils may serve both functions by modulating eicosanoid synthesis [10].

Breast cancer is the most common form of cancer among U.S. women. The lower incidence of breast cancer and other cancers among populations such as Greenland Eskimos, Japanese, and Icelanders that regularly consume fish [2,11,12], suggests that fish lipids, particularly n-3 PUFAs, may protect against the growth of breast cancer. When these populations have altered their diets to more western eating habits, mortality rates from cancer increased [12,13].

EXPERIMENTAL EVIDENCE

Mechanisms suggested for enhanced tumor growth by dietary fats include alterations of endocrine balance, stimulation of cell division and/or differentiation in mammary tissue, alterations in the immune competency of the host, and enhanced synthesis of prostaglandins (PGs), particularly PGE_2. PUFAs also may easily undergo oxidation to yield a variety of potential mutagens, promoters, and carcinogens.

PGE_2 suppresses immunological cells, which may facilitate tumor cell metastasis and proliferation [9,14]. There is a good correlation between the levels of PGE_2 and tumor growth in experimental animals [15]. Dietary n-6 PUFAs such as linoleic acid are elongated and desaturated to arachidonic acid, which is the precursor of PGE_2. Dietary n-3 PUFAs, which replace n-6 PUFAs in tumor tissue phospholipids as well as reduce tumor tissue synthesis of n-6 PUFAs, may exert an antitumor effect by reducing tumor PGE_2 synthesis.

Jurkowski and Cave [4] observed that diets containing 20% menhaden oil reduced tumor incidence and prolonged latent mammary tumor growth in rats compared to the enhanced tumor development and shortened latent period in rats receiving an equivalent dietary level of corn oil. The injection of n-3 PUFAs significantly depressed mammary tumor growth and simultaneously inhibited the synthesis of PGE_2 [3]. In animal studies dietary fish oil decreased the formation of PGE_2 and retarded the growth of tumor cells [3,16,17].

Dietary fish oil also caused a significant increase in n-3 PUFA levels in tumor lipids with a concommitant decrease in n-6 PUFAs

[4]. Dietary n-3 PUFAs apparently inhibit cyclooxygenase, thereby reducing the synthesis of PGE_2 [3]. This also diverts arachidonic acid into the lipoxygenase pathway, producing hydroxy fatty acids and leukotrienes that may enhance immunological cell activity and inhibit tumor cell growth.

Massive invasions of macrophages have been observed in tumors. Macrophages, which are major producers of prostaglandins, can affect tumor growth in two ways. First, because they are cytotoxic to tumor cells, macrophages may directly inhibit tumor growth. Second, by secreting PGE_2 and thus suppressing the immune antitumor response, macrophages may promote tumor growth [29]. In addition, macrophage PGE_2 has been shown to enhance tumor dissemination [30].

With high levels of PGE_2, macrophages do not function in their normal cytotoxic capacity against tumors [18,19]. By reducing the local concentration of PGE_2, macrophage antitumor function may be restored. Dietary intervention with fish oils may provide such an approach.

Macrophages effectively take up dietary n-3 PUFAs, which reduce macrophage arachidonic acid levels and subsequently decrease their capacity to synthesize PGE_2 [3,4,20]. By decreasing the overall PGE_2 levels and thus relieving inhibition of macrophage antitumor activity, dietary n-3 PUFAs may retard the growth of tumor cells [3].

In addition, it has been observed that dietary n-3 PUFAs enhance macrophage production of arginase, an enzyme that is cytolytic to tumors [21]. Thus dietary n-3 PUFAs could retard tumor growth without affecting normal macrophage function.

Inhibition of prostaglandin synthesis may enhance the effectiveness of certain antitumor agents. Some leukemic cells are made more responsive to chemotherapeutic drugs by n-3 PUFA uptake [22]. n-3 PUFAs also enhance the susceptibility of tumor cells to hyperthermia [23]. Thus a combination of nutritional intervention and chemotherapy may facilitate cancer treatment without severely affecting normal cell function.

Prostaglandin production peaks in the early stages of cancer development, and the highest prostaglandin levels have been observed in metastasizing and invasive tumors [24]. Hypercalcemia associated with cancer has been attributed to the overproduction of prostaglandins by tumors. Dietary menhaden oil significantly lowered prostaglandin production in mice with fibrosarcoma tumors and reduced plasma calcium concentration. However, tumor growth was not inhibited [25].

DISCUSSION

Current cancer research is focusing on agents that inhibit tumor prostaglandin synthesis. High rates of prostaglandin synthesis in tumor tissue may or may not indicate a causal relationship to tumor growth. However, the growth of many experimental tumors has been inhibited by inhibitors of prostaglandin synthesis, such as indomethacin. It also has been shown that the sensitivity of tumor cells to destruction by lymphocytes and phagocytes is increased when prostaglandin synthesis is inhibited [26].

In view of the potent effects of prostaglandins it would seem prudent to develop dietary treatments for the effective reduction of prostaglandin precursors, and in this regard, to investigate the effectiveness of dietary fish oils and/or fish lipids. Evidence that dietary n-3 PUFAs can reduce arachidonic acid concentrations and also inhibit prostaglandin synthesis is extremely significant in this context and deserves systematic study [27,28]. Dietary n-3 PUFAs may therefore be useful in reducing risk from cancer as well as in reducing coronary disease.

REFERENCES

1. Ames, B.N. (1983). Dietary carcinogens and anticarcinogens. Science 221:1256-63.
2. Nielsen, N.H. and J.P. Hansen (1980). Breast cancer in Greenland; selected epidemiological, clinical, and histological features. J. Cancer Res. Clin. Oncol. 98:287-99.
3. Karmali, R.A., J. Marsh, and C. Fuchs (1984). Effect of n-3 fatty acids on growth of a rat mammary tumor. J. Natl. Cancer Inst. 73:457-61.

4. Jurkowski, J.J. and W.T. Cave (1985). Dietary effects of menhaden oil on the growth and membrane lipid composition of rat mammary tumors. J. Natl. Cancer Inst. 74:1145-50.

5. Gabor, H., L.A. Hillyard, and S. Abraham (1985). Effect of dietary fat on growth kinetics of transplantable mammary adenocarcinoma in BALB/c mice. J. Natl. Cancer Inst. 74(6): 1299-305.

6. Carrol, K.K. and G.J. Hopkins (1979). Dietary polyunsaturated fat versus saturated fat in relation to mammary carcinogenesis. Lipids 14:155-58.

7. Carrol, K.K. and H.T. Khor (1975). Dietary fat in relation to tumorigenesis. Prog. Biochem. Pharmacol. 10:308-53.

8. Hillyard, L.A. and S. Abraham (1979). Effect of dietary polyunsaturated fatty acids in growth of mammary adenocarcinomas in mice and rats. Cancer Res. 39:4430-37.

9. Plescia, O.J. (1982). Does prostaglandin symthesis affect in vivo tumor growth by altering tumor/host balance? In Prostaglandins and Cancer: First International Conference, pp.619-31. T.J. Powles, R.S. Bockman, K.V. Honn, and P. Ramwell, (eds). A.R. Liss, Inc., New York.

10. Bang, H.O., J. Dyerberg, and N. Hyorne (1976). The composition of food consumed by Greenland Eskimos. Acta. Med. Scand. 220:69-73.

11. Munro, I. (1983). Eskimo diets and disease. Lancet I,1139-41.

12. Miller, A.B. (1980). Nutrition and cancer. Prev. Med. 9:189-96.

13. Kagawa, Y. (1978). Impact of westernization on the nutrition of Japanese: changes in physique, cancer, longevity, and centenarians. Prev. Med. 7:205-17.

14. Plescia, O.J., A.H. Smith, and K. Grinwick (1975). Subversion of immune system by tumor cells and role of prostaglandins. Proc. Natl. Acad. Sci. 72:1848-54.

15. Fulton, A., A. Rios, S. Loveless, and G. Heppner (1982). Prostaglandins in tumor associated cells. In Prostaglandins and Cancer: First International Conference, pp.701-03. T.J. Powles, R.S. Bockman, K.V. Honn, and P. Ramwell, (eds). A.R.Liss Inc., New York.

16. Karmali, R.A., S. Wilt, H.T. Thaler, and F. Lefevre (1983). Prostaglandins in breast cancer. Relationship to disease stage and hormone states. Br. J. Cancer 48:689-96.

17. Hopkins, G.J., T.G. Kennedy, and K.K. Carroll (1981). Polyunsaturated fatty acids as promoters of mammary carcinogenesis induced in Sprague Dawley rats by 7,12-dimethylbenz-(a)anthracene. J. Natl. Cancer Inst. 66:517-22.

18. Goodwin, J.S. and J. Ceuppens (1983). Regulation of the immune response by prostaglandins. J. Clin. Immunol. 3:295-315.

19. Schultz, R.M., N.A. Pavlidis, W.A. Stylos, and M.A. Chirigos (1978). Regulation of macrophage tumoricidal function: a role for prostaglandins of the E series. Science 202:320-27.

20. Magrum, L.J. and P.V. Johnston (1983). Modulation of prostaglandin synthesis in rat peritoneal macrophages with n-3 fatty acids. Lipids 18:514-21.

21. Johnston, P.V. and L.A. Marshall (1984). Dietary fat, prostaglandins, and the immune response. Prog. Food Nutr. Sci. 8:3-25.

22. Guffy, M.M., J.A. North, and C.P. Burns (1984). Effect of cellular fatty acid alteration on adriamycin sensitivity in cultured L1210 murine leukemia cells. Cancer Res. 44:1863-66.

23. Guffy, M.M., J.A. Rosenberger, I. Simon, and C.P. Burns (1982). Effect of cellular fatty acid alteration on hyperthermic sensitivity in cultured L1210 murine leukemia cells. Cancer Res. 42:3625-30.

24. Smith, B.J., M.R. Wills, and J. Savory (1983). Prostaglandins in cancer. Annals Lab. Sci. 13:359-65.

25. Tashjian, A.H., E. Voelkel, D.R. Robinson, and L. Levine (1984). Dietary menhaden oil lowers plasma prostaglandin and calcium in mice bearing the HSDM fibrosarcoma. J.Clin.Invest. 74:2042-48.

26. Fulton, A.M. and G.H. Heppner (1985). Relationships of prostaglandin E and natural killer sensitivity to metastatic potential in murine mammary adenocarcinomas. Cancer Res. 45:4779-84.

27. Karmali, R.A. (1980). Review: Prostaglandins and cancer. Prostaglandins Leukotrienes Med. 5:11-28.

28. Levine, L. (1981). Arachidonic acid transformation and tumor production. Adv. Cancer Res. 3549-79.

29. Goodwin, J.S. and J.L. Ceuppens (1985). Prostaglandins, cellular immunity and cancer. In Prostaglandins and Immunity, pp 1-34. J.S. Goodwin (ed), Martinus Nijhoff Publishing, Boston.

30. Young, M.R. and M. Newby (1986). Enhancement of Lewis lung carcinoma cell migration by prostaglandin E2 produced by macrophages. Cancer Res. 46:160-64.

6
Fish and n-3 Polyunsaturated Fatty Acid Consumption in the United States: Some Calculations

INTRODUCTION

The per capita consumption of fish and shellfish in the United States averaged 12.8 pounds per annum during the years 1979 to 1983 and is currently around 14 pounds [1]. In this chapter, estimates of n-3 polyunsaturated fatty acid (n-3 PUFA) intake at this and other levels of seafood consumption are made. The resulting values are compared to the n-3 PUFA intake of Eskimos, known to have a diet rich in seafood, and to the levels provided in human feeding trials using fatty fish (see Chapter 3).

THE n-3 PUFA CONTENT OF FISH

Various species of salmon, tuna, mackerel, and trout are among the fish with fatty flesh readily available in U.S. markets (see Chapter 12). The EPA (eicosapentaenoic acid, 20:5 n-3) and DHA (docosahexaenoic acid, 22:6 n-3) contents of fillets from these fish are presented in Table 6.1. From these values, the average content of EPA and DHA in fatty fish fillets is calculated to be 0.68 g and 1.0 g/100g fillet, respectively.

Fish with lean flesh available to American customers include striped bass, cod, flounder, haddock, and sole. The EPA and DHA contents of fillets from these fish are shown in Table 6.2. The average EPA and DHA levels in these fillets are 0.1 g and 0.17 g/100 g, respectively.

171

TABLE 6.1 n-3 Fatty Acid Content in Fillets of Fatty Fish Available in U.S. Markets

Species	EPA g/100g fillet	DHA g/100g fillet
Mackerel, Atlantic	0.65	1.10
Mackerel, Pacific (eviscerated)	1.10	1.30
Salmon, Atlantic	0.18	0.61
Salmon, Chinook	1.00	0.72
Salmon, chum	0.24	0.31
Salmon, coho	0.82	0.94
Salmon, pink	0.64	0.86
Salmon, red	1.30	1.70
Trout, rainbow	0.22	0.62
Tuna, albacore (white meat)	0.63	1.70
Average	0.68	1.00

Source: Ref. 4

n-3 PUFA INTAKE WITH THE CURRENT U.S. SEAFOOD CONSUMPTION

Of the 14 lbs. of fish and shellfish consumed annually by the American population, 1 to 2 lbs. are shrimp [1]. Thus approximately 12 to 13 lbs. of finfish are consumed per annum, corresponding to 0.031 lbs. or 14 g/day. If it is assumed that all fish is consumed as fatty fish fillets, the average daily consumption is 0.095 g of EPA and 0.14 g of DHA. If half the fish consumed are lean species, the daily per capita intake would be 0.055 g of EPA and 0.082 g of DHA.

TABLE 6.2 n-3 Fatty Acid Content in Fillets of Lean Fish Available in U.S. Markets

Species	EPA g/100g fillet	DHA g/100g fillet
Bass, striped	0.17	0.47
Cod, Atlantic	0.08	0.15
Cod, Pacific	0.07	0.12
Flounder, yellowtail	0.11	0.11
Haddock	0.05	0.10
Sole, lemon	0.09	0.09
Average	0.10	0.17

Source: Ref. 4

n-3 PUFA INTAKE WITH A HYPOTHETICAL HIGHER CONSUMPTION

If the U.S. population consumed 225 g of seafood a day (two daily servings of approximately one-fourth pound each) in the form of fatty fish fillets, a daily intake of 1.53 g of EPA and 2.5 g of DHA would be obtained. If this dietary pattern was followed by the entire U.S. population (approximately 235 million), 42,458,625,000 lbs. of fish fillets (less than 50% of whole fish) would be consumed yearly. In comparison, the 1983 total U.S. commercial fish catch was only 5,539,463,000 lbs. [1].

A more realistic goal for seafood consumption by the U.S. population would be consumption of four four-ounce servings a week, which would amount to 0.142 lbs. or 65 g of fish a day. If these servings were fatty fish fillets, 0.44 g of EPA and 0.65 g of DHA would be consumed daily. If followed by the entire population, this pattern of fish consumption would utilize 1,218,005,000 lbs. per year, about 20% of the total U.S. commercial fish catch in 1983.

COMPARISON TO THE n-3 PUFA INTAKE OF ESKIMOS AND FEEDING TRIAL PARTICIPANTS

Greenland Eskimos consuming a diet composed predominantly of seafood consume approximately 5.7 g of EPA/day [2]. The U.S. population intake is much lower (Table 6.3). Even if seafood consumption reached 225 g/day and this intake was in the form of fatty fish, the EPA intake would be only slightly above 1.5 g/day.

In contrast, volunteers in feeding trials using fatty fish consumed EPA in amounts ranging from 1 g to 15 g/day (see Chapter 3). Of the feeding trials listed in Table 3.1, which presents values for daily EPA and DHA consumption, five studies provided EPA in quantities greater than 1.5 g, and in two studies consumption was greater than 5.7 g. The DHA intake was included in one of the feeding studies, and at 2.8 g it also was greater than what could be realistically consumed by Americans.

To have EPA consumption similar to that in the feeding trials, extremely large quantities of seafood would need to be consumed. The consumption of 5 g of EPA/day requires 735 g of fatty fish fillets;

TABLE 6.3 Estimated Approximate Intake (g/day) of Unsaturated Fatty Acids by Different Population Groups

	18:2n-6	18:3n-3	20:5n-3	22:6n-3
U.S.	8 - 20	0.5	---	trace
Japanese	2 - 5	0.4	---	1 - 3
Eskimo	2 - 3	---	---	6 - 12

to consume 10 g of the same fatty acid, almost 1.5 kg of fillets need to be eaten!

The results of many of the fatty fish feeding trials are thus of questionable practical significance because of the exceedingly large quantities of fatty fish used. Feeding trials in which the seafood intakes are at levels that could be followed realistically by the American population are needed. A questionable epidemiological study among Dutch men suggested that as little as one or two servings of seafood per week may prevent cardiovascular disease [3]. If the beneficial cardiovascular changes observed in feeding trials in which much greater n-3 PUFA levels were consumed are also evident with smaller intakes, changing the American diet to regularly include more seafood should perhaps become a goal of nutrition education programs.

Consuming fish oil at 15 to 30 g/day may not be generally practicable even with an adequate supply of refined edible fish oil. However, the effective doseage of n-3 PUFA is greatly determined by the amount and type of dietary fat.

The high intake of linoleic acid (15 to 20 g/day) means that tissue concentrations of linoleic acid in Americans are high, and therefore, large amounts of n-3 PUFAs would be required consistently over a long period to dilute tissue n-6 pools and perhaps to exert their effects on eicosanoid production. Hence, achieving the prophylactic effect of dietary fish oil may require adjustment of dietary fat composition and intake as part of a total dietary regimen, in which fat may be about 20 to 25 en% and composed of saturated (25%), monoenoic (40%), dienoic (18:2n-6, 20%), and polyenoic (n-3, 15 to 20%) fatty acids. With a reduced intake of total fat and n-6 PUFAs, the effective doseage of n-3 PUFAs may be attainable and feasible with reasonable seafood consumption.

REFERENCES

1. National Marine Fisheries Service (1984). <u>Fisheries of the United States, 1983.</u> Current Fishery Statistics #8320.

2. Dyerberg, J. (1982). Observations on Populations in Greenland and Denmark. In <u>Nutritional Evaluation of Long Chain Fatty Acids in Fish Oil</u>, S.M. Barlow and M.E. Stansby (eds.), p. 245-261. Academic Press, London.

3. Kromhout, D., E.B. Bosschieter, C de L. Coulander (1985). The inverse relation between fish consumption and 20-year mortality from coronary heart disease. <u>N. Eng. J. Med.</u> 312(19):1205-1209.

4. Exler, J. and J.L. Weihrauch (1976). Comprehensive evaluation of fatty acids in foods. VIII. Finfish. <u>J. Am. Diet. Assoc.</u> 69:23-48.

7
Cholesterol and Fat Soluble Vitamins in Fish Lipids

CHOLESTEROL CONTENT OF FISH OILS

Sterols, along with vitamins A, D, and E, are the major components of the unsaponifiable portion of fish oils, and cholesterol is the most common sterol in most marine species [1].

While most fish oils contain small amounts of unsaponifiable matter, the liver oil of elasmobranchs contains high concentrations, of which cholesterol is a component. When the unsaponifiable fraction of an elasmobranch liver oil is low (i.e. 1% to 2%), it is predominantly cholesterol; when 10% to 35%, it contains cholesterol and other saturated and unsaturated long-chain alcohols; when high, the unsaponifiable fraction consists mostly of squalene [1,2].

In contrast to elasmobranchs, cholesterol is one of the major constituents of the unsaponifiable fraction of teleost body oils [2]. Oils extracted from red muscle [2] or dark flesh [3] have higher amounts of cholesterol. The literature contains limited data on the cholesterol content of fish oils. Peifer [4] reported the sterol concentrations of menhaden (480 mg/100g oil), mullet (2,400 mg/100g oil), ocean perch (2,600 mg/100g oil), and silver salmon (810 mg/100g oil).

Kristjanson [5] reported that halibut liver oil contained 7.74% cholesterol. Wurzinger [3] found 0.15% to 0.52% cholesterol in cod liver oil. Grace and Sen [6] reported an average cholesterol content of 1% in sardine oil. Jacquot [2] reported 0.3% cholesterol in the body oil of the arctic stickleback.

The cholesterol content of fish oils and the edible portion of raw flesh for finfish, crustaceans, and mollusks are presented in table 7.1. Cod liver, herring, menhaden, and salmon oils range from 485 mg to 766 mg/100g oil.

Due to the current demand for information on the content of lipid components in seafoods, more data are available on the cholesterol content of fish flesh than for fish oils. Many people still assume that fish products are high in cholesterol. However, Ackman [7] calculated that one 100-gram serving of fish contributes only one-tenth of the daily cholesterol intake in a diet containing 0.5% cholesterol, corresponding to an intake of 300 mg to 500 mg/day. The flesh of finfish contains less than 100 mg cholesterol/100 g edible portion, raw, while crustaceans and mollusks may contain higher amounts of total sterols (Table 7.1). Although shellfish have the reputation of being high-cholesterol foods, in several cases those sterol values include sterols other than cholesterol [7].

VITAMIN A AND D CONTENT OF FISH OILS

Until vitamins A and D were synthesized and produced commercially, marine oils were the most important natural source of these vitamins. The body and liver oils of fish are particularly rich in vitamins A and D and still are used in animal feed. The livers of many fish species contain relatively large concentrations of oil-soluble vitamins, which are associated with the lipid fraction. The amounts vary widely both within and between species, and the vitamin content of liver oils varies with species, age, size, sex, nutritional condition, and spawning stage of the fish, as well as the geographical locale and season of catch [8]. Variations in feeding and the reproductive cycle are most likely responsible for seasonal differences observed in vitamin content of fish oils [9].

Vitamin A

Fish derive vitamin A from their diet. The biologically active forms of vitamin A in fish oils are retinol (vitamin A1) and 3-dehydroretinol (vitamin A2). In marine fish, retinol is the major

TABLE 7.1 The Cholesterol Content of Fish Tissue and Oil

Fish Tissue	Cholesterol mg/100g	Reference
Bass, freshwater	59.0	18
Bass, striped	80.0	18
Bluefish	59.0	18
Burbot	59.5	18
Carp	67.2	13
Catfish	30.2-56.0	7
Channel	58.0	18
Brown bullhead	75.0	18
Cod, European	41.2	7
North American	50.0	19
Atlantic	43.0	18
Pacific	37.0	18
Canadian	26.0	18
Codling, European	22.0-35.0	7
Croaker, Atlantic	61.0	18
Dogfish, European	28.4-73.0	7
Spiny	51.8	18
Freshwater drum	64.0	18
Flounder, North American	50.0	19
		46.218
Jewfish grouper	49.0	18
Haddock, N. American	90.0	7
	62.9	18
European	27.0-35.0	7
unspecified	45.0-90.0	21
Halibut, black European	43.0-50.0	7
White European	24.0-34.0	7
North American	60.0	19
Pacific	32.0	18
Greenland	46.0	18

TABLE 7.1 Cholesterol Content of Fish Tissue and Oil (continued)

Fish Tissue	Cholesterol mg/100g	Reference
Herring, European	53.2-66.4	7
North American	53.2-66.4	19
Atlantic	60.0	18
Pacific	76.6	18
Ling, European	29.0-46.0	7
Mackerel, European	34.0-38.0	7
North American	95.0	19
Atlantic	80.0	18
chub	52.0	18
horse	41.0	18
Japanese horse	48.0	18
king	53.0	18
Mullet, striped	49.0	18
Perch, red, European	21.3-54.4	7
North American	50.0-70.0	19
ocean	42.0	18
white	80.0	18
yellow	90.0	18
Pike, northern	39.0	18
walleye	86.0	18
Plaice, European	70.0	18
Pollack, European	31.0-36.0	7
North American	70.6	18
Pompano, Florida	50.0	18
Rockfish	49.0	18
Salmon, European	53.0	7
North American	95.0	7
Chum	74.0	18
Pink	65.0	21
Sockeye	19.0	18
Smelt, rainbow	70.0	18
Sole, European	50.5	18
North American	50.0-70.0	19

TABLE 7.1 Cholesterol Content of Fish Tissue and Oil (continued)

Fish Tissue	Cholesterol mg/100g	References
Sprat	38.5	18
Sucker, white	43.0	18
Sunfish, pumpkinseed	67.0	18
Swordfish	39.0	18
Trout, North American	55.0	19
brook	68.0	18
lake	48.0	18
rainbow	57.0	18
Tuna, North American	65.0	19
albacore	54.0	18
bluefin	38.0	18
skipjack	47.0	18
Whitefish, lake	60.0	18
Whiting, North American	50.0-70.0	19
	75.0	21
	31.0	18

<u>Marine Invertebrates</u>

Crustaceans		
Crab, blue	78.0	18
	70.0-98.0	21
Dungeness	52.0-63.0	21
	59.0	18
queen	127.0	18
Crayfish, other	518.0	18
Lobster, European	129.0	18
northern	95.0	18
unspecified	170.0-350.0	21
Shrimp, Atlantic brown	142.0	18
Atlantic white	182.0	18
Japanese prawn	58.0	18
northern	125.0	18
unspecified	140.0	18
	138.0-200.0	21

TABLE 7.1 (continued)

Fish Oil	Cholesterol mg/100g	Reference
Mollusks		
Conch, other	141.0	18
Mussel, blue	38.0	18
Oyster, blue point	37.0-58.0	21
eastern	47.0	18
European	30.0	21
unspecified	112.0-470.0	21
Periwinkle, common	101.0	18
Scallop, Atlantic deepsea	37.0	18
unspecified	60.0-175.0	21
Cod liver	570.0	18
Herring	766.0	18
Menhaden	521.0	18
Salmon	485.0	18
MaxEPA*	600.0	18

*concentrated fish body oils, registered tradename

form, although 4% to 20% of the total vitamin may be dehydroretinol [10]. In freshwater fish, dehydroretinol is more predominant [9]. Approximate ratios between retinol and dehydroretinol were 2:1 and 1:2, respectively, in marine and freshwater fish liver samples [11]. High performance liquid chromatography of marine samples has allowed the identification of as many as 13 geometric isomers of retinol and dehydroretinol, and vitamin A biopotency varies with the relative distribution of these isomers [11].

Marine oils vary markedly in vitamin A content even within species (Table 7.2). Vitamin A is concentrated in the fish liver and is accumulated as the fish ages. In mature fish approximately 90% of the total vitamin A in the body is stored in the liver [10]. In some species the vitamin A potency of the liver oil is due to the increase

TABLE 7.2 The Vitamin A Content of Fish Liver Oils*

Fish, Marine	Vitamin A (I.U./gram oil)
Cod	1,000- 17,000
Codling	100- 300
Haddock	100- 3,000
Hake	1,000- 25,000
Halibut	5,000-330,000
Herring, Pacific	30,000- 60,000
Ling	500

Mackerel, California
  Chub
  Horse
	45,000
	1,400-420,000
	69,000-512,000

Plaice, Northeast Atlantic	3,000
Pilchard, Pacific	30,000- 65,000
Pollock Alaska	2,000 500-130,000
Rockfish	13,000-320,000

Salmon, Atlantic
  Chinook
  Chum
  Pink
  Sockeye
	5,000- 20,000
	25,000- 40,000
	5,000- 15,000
	10,000- 40,000
	10,000- 50,000

Sardine	16,000
Seabass	40,000-500,000

Shark, dogfish
  great blue
  hammerhead
  soupfin
	200-200,000
	100- 43,000
	30,000-120,000
	10,000-200,000

Sturgeon, white	10,000- 20,000
Swordfish	20,000-400,000

TABLE 7.2 (continued)

Fish, Marine	Vitamin A (I.U./gram oil)
Tuna, albacore	10,000- 60,000
bluefin	25,000-100,000
Japanese	170,000
skipjack	30,000- 60,000
striped	36,000
yellowfin	35,000- 90,000
yellowtail	20,000-200,000
Turbot	10,000
Whiting	200,000
Fish, Freshwater	
Burbot	4,000
Carp	5,000- 7,000
Perch	10,000
Pike	3,000- 14,400
Sturgeon	26,000
Whitefish	59,500

*Average ranges from reviews by Cruickshank (9) and Higashi (10)

in size with age, while in other species age alone determines the vitamin potency [9].

Fish species have been categorized on the basis of oil content and vitamin potency. Fish with a high oil content and low vitamin A potency include the cod and dogfish shark; halibut, tuna, and whale have a low oil content and high vitamin A potency; high vitamin A potency and high oil content are characteristic of the soupfin and other shark species [8].

Elasmobranchs, or cartilaginous fish, accumulate vitamin A in the liver and much less in other body tissues [9]. In teleosts,

TABLE 7.3 The Vitamin D Content of Fish Liver Oils

Fish, Marine	Vitamin D (I.U./gram oil)
Cod	60- 300
Atlantic	20- 300
gray	85- 500
Pacific	85- 500
Red	300- 5,000
Flounder, starry	1,000
Haddock	50- 100
Hake, squirrel	120
Halibut, Atlantic	550-20,000
Greenland	400
Pacific	1,000- 5,000
Herring, Pacific	250
Ling	100- 500
Mackerel, Atlantic	750- 6,000
Pacific	1,400- 6,300
Pilchard, Pacific	2,300
Pollock	20- 70
Salmon, Atlantic	180
pink	100- 600
spring	100- 600
Shark, dogfish	5- 30
soupfin	5- 25
Swordfish	2,000-25,000
Tuna, albacore	13,000-250,000
bluefin, Atlantic	16,000- 30,000
Pacific	20,000- 70,000
oriental	45,000
skipjack	25,000-250,000
striped	42,000
yellowfin	10,000- 45,000
Whiting	1,000

TABLE 7.3 (continued)

Fish, Freshwater	Vitamin D (I.U./gram oil)
Carp	10,000
Ling	400
Muskellunge	500
Sturgeon	<1

Source: Refs. 9, 22, and 12.

vitamin A accumulates in the liver as well as in other body tissues. Oil from the alimentary tract of certain fish contains large amounts of vitamin A (2,000 I.U to more than 500,000 I.U./g oil)[9]. The flesh of fatty fish is rich in vitamin A, while the flesh of lean fish is a poor source [2].

Vitamin D

The predominant form of vitamin D in fish oils is vitamin D3 [12]. The fish liver is the richest source of vitamin D. Although vitamin D is concentrated in fish liver oils, most oils contain only moderate amounts. Some tuna species have the highest concentration of vitamin D, with up to 250,000 I.U./g liver oil [12]. Unlike vitamin A, vitamin D is not present in nonliver visceral oils in amounts exceeding liver oil content [9].

The variation of vitamin D content between and within species is much less than that of vitamin A [10]. Freshwater fish oils contain much lower amounts of vitamin D than oils from marine fish (Table 7.3). Elasmobranch liver oils have low levels of vitamin D, in contrast to their relatively high vitamin A content.

VITAMIN E CONTENT OF FISH OILS

Since fish do not synthesize vitamin E, the measurable levels of tocopherol in fish oil are directly related to the fish's diet of algae and seaweed [13]. Tocopherol content of fish oil varies greatly

TABLE 7.4 The Tocopherol Content of Some Fish Oils

Fish Species	Tocopherol Content mg/100 grams oil	Reference
Anchovy, commercial	29.08	14
noncommercial	74.55	14
Capelin, commercial	14.00	14
Catfish, liver	110.00	15
Cod, Atlantic, commercial, liver	21.96	14
liver	11.50	15
liver	17.3	23
Haddock, liver	9.00	15
noncommercial	0.60	14
liver	18.00	14
Hake, Atlantic, liver	20.00	15
Herring, muscle	14.00	15
commercial	9.22	14
Ling cod, liver	9.00	15
Menhaden, muscle	7.00	15
commercial	7.50	14
Ocean perch, noncommercial	18.74*	14
Sable, muscle	63.00	15
Salmon, Chinook, noncommercial	19.15	14
pink, muscle	22.00	15
Sardine, muscle	4.00	15
Shark, dogfish, liver, commercial	25.00	14
Greenland, commercial	50.00	14
Skate, liver	22.00	15
Sole, lemon, liver	54.00	15
Tuna, muscle	16.00	15

TABLE 7.4 (continued)

Fish Species	Tocopherol Content mg/100 grams oil	Reference
Tuna, skipjack, liver, noncommercial	90.00*	14
Trout, rainbow, liver, noncommercial	5.20*	14
meat and skin, noncommercial	12.50*	14
Whale, muscle	20.00	15
Wolffish, Atlantic, liver		
noncommercial	185.50	14
meat, noncommercial	35.50	14
Wrasse, European, liver,		
noncommercial	250.67	14
Shellfish		
limpet	150.00	14
prawn	95.00	14
squid, commercial	21.00	14

* measured as total vitamin E

among and within fish species, and in almost all species it is higher in fish liver oil than in body oil [14]. Alpha-tocopherol is the most abundant form of vitamin E in fish oils [13].

The content of tocopherol in fish body oil ranges from less than 1 mg/100g oil to 75 mg/100g oil [14]. Noncommercial liver oil has greater amounts of tocopherol than oil that has been processed and used for commercial purposes (Table 7.4). Olcott [15] reported the tocopherol content of several fish oils (Table 7.4), but these should be considered as rough estimates because of variations due to the season of the catch and the diet of the fish.

Fish flesh is a good source of vitamin E. The tocopherol content is directly related to fat content. Syvaoja et al. [16] reported average tocopherol levels near 1.0 mg/100g in high-fat species (salmon and rainbow trout). When they calculated tocopherol as a component of fat, however, low-fat fish contained 1.1 mg/g fat, medium-fat fish (bream, whitefish, Baltic herring, and vandace) had 0.3 mg/g fat, and high-fat species contained 0.16 mg tocopherol/g fat.

The tocopherol content of fish flesh is related to geographic locale, sexual maturity, and seasonal factors, which include diet and temperature [16]. Tocopherol levels in tissues peak during spawning season [7, 16]. The dark tissue of cod contains more tocopherol than the white tissue [13]. Marine fish have higher levels of alpha-tocopherol than freshwater fish of the same species [16].

Some potential deleterious problems attributed to fish oils are related to their oxidation potential and a subsequent increase in peroxides [17]. The high degree of polyunsaturation in fish oil fatty acids may cause the accumulation of free radicals and potentially toxic levels of peroxides. Since tocopherols are naturally occurring antioxidants, their presence may enhance the stability of fish oils [15]. Ackman [7] showed that constituent tocopherol can protect living fish from free radical peroxidation, but that residual tocopherol may not provide adequate protection from post mortem autoxidation in frozen muscle. Although fish oils are comparatively rich in tocopherol, the antioxidant potency of tocopherol is limited in vitro [7].

REFERENCES

1. Stansby, M.D. (1982). Properties of fish oils and their application to handling of fish and to nutritional and industrial use. In Chemistry and Biochemistry of Marine Food Products, R.E. Martin, G.J. Flick, C.E. Hebard, D.R. Ward (eds). p. 75-92. Avi Publishing Co., Westport, CT.

2. Jacquot, R. (1961). Organic constituents of fish and other aquatic animal foods. In Fish as Food, Vol.1. G. Borgstrom (ed). p.146-209. Academic Press, New York.

3. Wurziger, J. (1977). Cholesterol content of cod liver oil. Fleischwirtschaft 57(9):1637-39

4. Peifer, J.J., F. Janssen, R. Muesing, and W.O. Lundberg (1962) The lipid depressant activities of whole fish and their component oils. J. Amer. Oil. Chem. Soc. 39:292-96.

5. Kristjanson, S. (1951). Separation of cholesterol from halibut liver oil. Fisheries Res. Bd. Can., Progress Rpts. Pacific Coast Stas. 88:51-52.

6. Grace, G., and D.P. Sen (1978). Cholesterol content of sardine oil. J. Oil. Tech. Assoc. India 10(3):137-38.

7. Ackman, R.G. (1974). Marine lipids and fatty acids in human nutrition. In Fishery Products, R. Kreuzer (ed) p.112-131. Fishing News (Books) Ltd., Surrey, England.

8. Karrick, N.L. (1969). Vitamins from marine sources. In The Encyclopedia of Marine Resources, F.E. Firth (ed).p.718-721. Van Nostrand-Reinhold Co., New York.

9. Cruickshank, E.M. (1962). Fat soluble vitamins. In Fish as Food, Vol.2, G. Borgstrom (ed). p.175-203. AcademicPress, New York.

10. Higashi, H. (1961). Vitamins in fish, with special referenceto edible parts. In Fish as Food, Vol.1, G. Borgstrom (ed). p.411-486. Academic Press, New York.

11. Stancher, B., and F.Zonta (1984). High performance liquid chromatography of the unsaponifiable from samples of marine and freshwater fish: fractionation and identification of retinol (vitamin A1) and 3-dehydroretinol (vitamin A2) isomers. J. Chromatog. 287:353-64.

12. Takeuchi, A., et al. (1984). High performance liquid chromatography determination of vitamin D3 in fish liver oils. J. Nutr. Sci. Vitaminol. 30(5):421-30.

13. Ackman, R.G. and M.G. Cormier (1967). Alpha-tocopherol in some live-holding without food. J. Fish. Res. Bd. Can. 24(2):357-73.

14. McLaughlin, P.J. and Weihrauch, J.L. (1979). Vitamin E content of foods. J. Amer. Diet. Assoc. 75:647-65.

15. Olcott, H.S. (1967). Antioxidants. In Fish Oils. Their Chemistry, Technology, Stability, Nutritional Properties, and Uses, M.E. Stansby (ed). p. 164-170. Avi Publishing Co., Inc. Westport, CT.

16. Syvaoja, E.-L., K. Salminen, V. Piironen, P. Varo, O. Kerojoki, and P. Koivistoinen (1985). Tocopherols and tocotrienols in Finnish foods: fish and fish products. J. Amer. Oil. Chem. Soc. 62(8):1245-48.

17. Karrick, N.L. (1967). Nutritional value as animal feed. In Fish Oils. Their Chemistry, Technology, Stability, Nutritional Properties, and Uses, M.E. Stansby (ed). p.362-382. Avi Publishing Co., Inc., Westport, CT.

18. Exler, J. and J.L. Weihrauch (1985). Provisional table on the content of omega-3 fatty acids and other fat components of selected foods. USDA Human Nutrition Information Service, HNIS/PT-103.

19. Feeley, R.M.et al. (1972). Cholesterol content of foods. J. Amer. Diet. Assoc. 61:134-51.

20. Kovacs, M.I.P., R.G. Ackman, and P.J. Ke (1978). Important lipid components of some fishery-based convenience food products:fatty acids, sterols, and tocopherols. J. Canad. Diet. Assoc. 39(3): 178-83.

21. Sidwell, V.D., P.R. Foncannon, N.S. Moore, and J.C. Bonnet (1974). Composition of the edible portion of raw crustaceans, finfish, and mollusks. I. Protein, fat, moisture, ash, carbohydrate, energy value, and cholesterol. Marine Fisheries Rev. 36(3):21-35.

22. Sand, G. (1967). Fish oil industry in Europe. In Fish Oils. Their Chemistry, Technology, Stability, Nutritional Properties, and Uses, M.E. Stansby (ed). p.405-421. Avi Publishing Co., Inc., Westport, CT.

23. Stancher, B. and F. Zonta (1983). High performance liquid chromatography of fat-soluble vitamins. Simultaneous quantitative analysis of vitamins D2, D3, and E. Study of percentage recoveries from cod liver oil. J. Chromatog. 256:93-100.

8
Components Affecting the Safety of Fish Oils

INTRODUCTION

Fish oils are easily oxidized, and several compounds in oxidized oils have been associated with cellular damage and disease [4,8]. Following partial hydrogenation, fish oils may also contain isomers of eicosamonoenoic and docosamonoenoic acids (erucic acid), which may alter cardiac tissue of laboratory animals. In addition, fish oils may be contaminated with DDT and PCBs, substances banned in the United States because of their toxic effects on wildlife and humans. Finally, metals such as lead, mercury, and cadmium, with well-known toxic effects, may be present in fish oils. This chapter examines concerns about the safety of using fish oils for human consumption.

THE AUTOXIDATION OF FISH OILS

Autoxidation Reactions

Unsaturated fatty acids, in the presence of appropriate free radical initiators, react readily with atmospheric oxygen and undergo autoxidation [1,47]. The process begins with the loss of hydrogen radicals from methylene groups of the cis 1,4 pentadiene systems in the unsaturated fatty acid molecules and may be initiated by transition metals, ultraviolet light, enzymes, and/or heat. After an initial induction period, the rate of oxygen absorption increases exponentially [1,2]. The fatty acid free radicals are highly reactive and combine with oxygen to form peroxy radicals. These in turn abstract hydrogen from additional unsaturated fatty acid molecules to form fatty acid hydroperoxides, the primary products of autoxidation [1,3,4,5]. Hydroperoxides are very unstable and break down to produce many types of secondary products. They may cleave to form alkoxy and hydroxy free radicals. Alkoxy radicals react with other free radicals and molecules or undergo carbon-carbon cleavage to form aldehydes, ketones, alcohols, hydrocarbons, esters, furans, lactones, etc. [4,6]. Unsaturated hydroperoxides can further react with oxygen or condense into dimers and polymers that subsequently break down into many products [4]. The extent of oxidation is determined by measuring the level of hydroperoxides and is expressed as a peroxide value [7].

Deleterious Consequences of Autoxidation

The autoxidation of oils causes rancidity and the development of unpleasant flavors and odors in food [11]. Hydroperoxides are tasteless; however, their decomposition products, saturated and unsaturated aldehydes, ketones, acids, and other oxidative compounds impart repulsive flavors and odors [1]. Rancidity can be detected at an early stage of oxidation because the small molecular weight compounds formed possess very low odor thresholds; only a few ppm or ppb are necessary to impart an unacceptable odor and flavor to fish oils [11].

In vivo hydroperoxides and their products disrupt vital cell functions, causing cellular damage and disease. Fatty acid free radicals can produce polypeptide chain scission, induce polymerization, destroy labile amino acids, and inactivate enzymes [8]. Biological membranes can be altered by the oxidation of the component unsaturated fatty acids, resulting in increased permeability and the swelling and lysis of organelles and cells [4,5,8]. Membrane lipid deterioration and polymerization of peroxidized lipids cause the accumulation of lipofuscin pigments in cells and contribute to the aging process [4].

Lipid peroxides and free radicals are also implicated in the development of inflammation, cancer, and arteriosclerosis [4]. Peroxides of unsaturated fatty acids such as arachidonic acid, dihomo-γ-linoleic acid, linoleic acid, γ-linolenic acid, and

Components Affecting the Safety of Fish Oils

linolenic acid can inhibit prostacyclin synthetase. The inhibition of this enzyme may increase the formation of platelet thrombi and promote endothelial lesions [7]. Presumably, peroxides of n-3 PUFAs can exert similar deleterious effects.

Autoxidative Reactions in Fish Oils

Because of their high concentration of fatty acids with five and six double bonds, fish oils are highly susceptible to autoxidation. The course of oxidation is often quite different in extracted fish oils and in lipids in fish tissues [10]. The rate of fish oil oxidation is significantly different from that of other oils. The break in the induction curve is less sharp, and the beginning of the increase in peroxide number occurs sooner [10,13].

The autoxidation of fish oils is the most important cause of deterioration in quality [1]. Undesirable flavors and odors develop at very low peroxide values at an early stage of oxidation, even during the induction period [13,16].

Effects of Consumption of Autoxidized Fish Oils

Deleterious consequences from consuming oxidized fish oils, due to the possible toxic effects of peroxides, have been reported from animal feeding trials. Because these compounds have a strong local irritant action [14], animals fed oxidized fish oil showed inflammation of the intestinal tract, damage to the mucous membrane of the stomach and intestines, and severe ulcerations, as well as decreased cellular respiration and enzyme inhibition [14]. The consumption of oxidized fish oils also destroyed vitamin A and vitamins of the B group [14].

Rats fed a diet containing 5% highly oxidized cuttle fish oil (peroxide value = 240) showed abrupt decreases in body weight and died within a week [12]. The consumption of less oxidized oil (peroxide value = 30) was not followed by any noticeable toxic effects. Rats fed a diet containing 5% of this oil had weight gains similar to those on a control diet [12]. A feeding trial lasting through fifteen months and two generations of rats showed similar results; the animals fed diets containing 20 wt% of either fresh (peroxide value = 2) or oxidized (peroxide value = 49) fish oils showed no differences in weight gain, food efficiency, reproduction, or death rate. A histological examination of the organs also showed no differences [14].

Preventing the Autoxidation of Fish Oils

Lipid oxidation may be minimized by antioxidants. Several isomers of vitamin E present in vegetable oils act by donating hydrogen to free radicals, forming compounds that are unable to continue the oxidation reactions [11]. During processing, antioxidants such as BHT (butylated toluene) and BHA (butylated hydroxyanisole), which also donate hydrogen to free radicals, may be added to food products.

Fish oils usually have very low levels of natural antioxidants [2]. Antioxidant compounds added during processing to protect vegetable oils or other animal fats from oxidation sometimes have less of an effect in crude fish oils [10,16].

Inhibition of oxidation is achieved by storing fish oils at low temperatures, limiting their exposure to oxygen and their contact with metallic ions. For every 10°C rise in the storage temperature of fatty fish, the oxidation rate increases by a factor of two to three [3].

The difficulties of producing a high quality fish oil begin if fish are brought to the oil-rendering plant without refrigeration [13]. In addition, both fish liver oils and fish body oils are obtained from raw materials rich in heme pigments, which are very active catalysts of fat oxidation [1]. The quality of the oil is also affected by the equipment employed in processing. Easily corroded equipment can lead to contamination of the oil with metallic catalysts [1]. Packaging that protects the oil from oxidation until it is purchased and used by consumers is as important as adequate production conditions.

Autoxidation In Vivo and Vitamin E Requirements

As the polyunsaturated content of diet or tissue increases, tissues become more susceptible to lipid peroxidation. To prevent such damage, increased amounts of vitamin E, the most effective in vivo inhibitor of lipid oxidation, are required [17,18]. Feeding trials with human volunteers and laboratory animals have demonstrated that fish or fish oil consumption raises the proportion of n-3 poluynsaturated fatty acids in plasma, platelet, and erythrocyte membranes and in other tissues and organs throughout the body (see Chapters 3 and 4). An increase in the requirement for vitamin E accompanies these changes, and dietary supplementation with vitamin E generally has been practiced in feeding trials. The potential effects that the increased dietary intake of n-3 polyunsaturated fatty acids might have on the vitamin E status of the American population requires further study.

Conclusion

Compared to other edible oils, those produced from marine sources are most susceptible to deterioration through autoxidation. Concern for consumer safety requires that this be avoided. Production conditions must be very carefully controlled and protective packaging used. The effect of production practices on the market price of fish oils is an important factor determining the place fish oils will have in the American diet. How an increased intake of fish oil might affect the vitamin E status of consumers requires further study.

THE EFFECTS OF MONOENOIC FATTY ACIDS: ERUCIC ACID

Results of Early Feeding Trials with Rapeseed Oil

Research conducted in the 1960s showed that rats consuming diets containing high levels of high-erucic acid (22:1n-9) rapeseed oil developed a transient lipidosis, or accumulation of fat, in the heart muscle. Fat infiltration reached a peak after three to five days of feeding. The main cause appeared to be impaired fatty acid oxidation by the mitochondria. The accumulated lipids were almost exclusively triglycerides. With prolonged feeding, lipidosis gradually disappeared. However, cell destruction, local inflammatory reactions, and areas of fibrous scar tissue then became evident [19,20].

The erucic acid component was assumed to be the major factor responsible for the cardiac lesions because of the close correlation between erucic acid content and the severity of lesions [19,20].

Moreover, glyceryl trierucate caused long-term lesions comparable in incidence and severity to those of rapeseed oil [20].

Other feeding trials demonstrated notable differences between species in sensitivity to erucic acid. Pigs appeared to be more resistant to the accumulation of lipids in the heart than rats [19]. Additionally, the irreversible lesions attributable to erucic acid in rapeseed oil were not found in samples of pig cardiac tissue [20].

Concern over the safety of fish oils arose because some fish oils contain isomers of erucic acid. The dominant 22:1 isomer in marine fish is cetoleic acid (22:1n-11), which originates from fatty alcohols in copepods [23,24]. Fish such as capelin, mackerel, and herring, which feed preferentially on copepods, have higher levels of this fatty acid [23]. The level of 22:1 is extremely low in menhaden oil because as adults, menhaden feed by filtering phytoplankton [24]. Copepods with 22:1 alcohols do not occur in freshwater environments. Freshwater fish do not accumulate 22:1 fatty acids. Among the isomers present, 22:1n-9 tends to predominate [24].

It is now believed that erucic acid may not have all the deleterious consequences once attributed to it. Some cardiac changes have been observed when feeding fats and oils containing no 22:1 fatty acids [20,25]. The level of fat or other dietary components may be more important in determining the causes of lipidosis in cardiac tissue. Mild cardiac lipidosis is associated with high-fat diets [25]. Lipidosis in rats fed diets containing either of two varieties of margarine, corn oil, or low erucic acid rapeseed oil was not correlated to the quantities of 22:1 ingested [21].

The apparent susceptibility to cardiac lipidosis and lesions observed in rats during feeding trials may also be attributed to their greater intake of oil per unit of body weight compared to other experimental animals [26,27]. Rats have been fed fats and oils in amounts ranging from 0.7 to 3.9 g/kg body wt/day while pigs have consumed 0.5 g/kg body wt/ day [27]. To equal the intake of weanling rats, an average 60 kg human would have to consume 1.3 to 2.3 kg of fat a day, raising doubts that weanling rats should be regarded as a reference animal [25].

Components Affecting the Safety of Fish Oils 199

There is also some doubt that lipidosis per se is an abnormal or harmful physiological condition for rats. It has been observed to occur during suckling, with no apparent negative consequences in later life [21]. Mild to moderate cardiac lipidosis also is induced by a temporary (one or two day) lack of food, such as occurs in wild animals [25]. The changes in cardiac tissue reported in feeding trials have not been accompanied by severe cardiac insufficiency [20].

Changes in Human Cardiac Tissue

Autopsies have shown that 22:1 fatty acids often are present in heart tissues and serum lipids [25]. Mild lipidosis and necrosis are also frequent observations. These changes may be caused by a number of drugs, toxins, and clinical conditions [20]. Mild cardiac lipidosis may be widespread in humans consuming a western high-fat diet irrespective of the type of fat consumed [25]. Because of the heart's capacity to compensate for minor injury, performance may be unaltered [20]. During the decades that extracted fish oils have been a component of human diets, there has been no linkage between their consumption and heart disease [25].

Conclusion

The evidence to date suggests that the consumption of 22:1 fatty acids is probably not harmful to experimental animals or humans. The level of fat or other dietary components may play a greater role in inducing changes in cardiac tissue than monoenoic acids. Neither reports from rat feeding trials nor examinations of human hearts indicate these changes cause cardiac insufficiency. Moreover, epidemiological studies have not shown evidence that populations, such as Eskimos, with a relatively high intake of 22:1 fatty acids have more cardiac lesions [28,29].

THE PRESENCE OF ORGANOCHLORINE PESTICIDES AND POLYCHLORINATED BIPHENYLS

DDT [(2,2 - bis(p-chlorophenyl)-1,1,1-trichloroethane; also abbreviated as p,p'-DDT] was used as an insecticide in the 1940s [30]. Due to numerous reports of adverse effects on wildlife, its agricultural use was banned in the United States in 1972 [31,32]. Similar regulation is in effect in other industrialized nations [32]. However, DDT still is used extensively in developing nations [32].

Polychlorinated biphenyl compounds (PCBs) were introduced in the 1920s for various industrial uses [35]. In rats, regular PCB ingestion produces hepatic tumors. In Japan, many people became ill after accidentally ingesting rice oil containing 100 ppm of PCBs [33].

The occurrence and action of DDT, PCBs, and other organochlorine compounds in the environment depends to a large extent on three chemical and physical characteristics. First, at normal environmental temperatures these compounds have a vapor pressure that allows them to exist both as a free vapor and adsorbed to atmospheric particles. In either manner, they become widely dispersed in the atmosphere and return to land or aquatic environments in rainfall. Second, they are chemically stable and markedly resistant to biological and abiotic degradation. Finally, they are relatively insoluble in water and soluble in lipids. Because of these chemical and physical properties and their widespread use, DDT, its metabolites (p,p'-DDT and p,p'-DDE), and PCBs are ubiquitous constituents of aquatic environments, and they can be identified in aquatic organisms [30,32].

In the aquatic environment, organochlorine compounds usually exist in association with biota rather than in solution. Their concentration in an organism may depend on its position in the food web; organisms at lower trophic levels contain lower concentrations than those at higher levels. Some researchers believe that absorption of organochlorines from water is the more important uptake route and that this route is controlled by properties such as their lipid-water partition coefficient [30,32].

Organochlorine Pesticides and Polychlorinated Biphenyls in Fish

In addition to exposure, total body lipid content is an important parameter determining the level of organochlorines in fish. Being lipophilic, these compounds accumulate preferentially in fatty fish species [32]. Among fish caught in waters off the Canadian Maritime Provinces, alewives, dogfish, herring, mackerel, salmon, and bluefish tuna--all fatty pelagic fish--had PCB and DDT (DDT = [p,p'-DDT + p,p'-DDD + p,p'-DDE]) levels above 0.1 ug/g. The greatest mean concentrations of PCB and DDT were found in bluefin tuna and were 3.9 ug/g and 3.1 ug/g, respectively. In comparison, groundfish, such as cod, haddock, plaice redfish, and yellowtail flounder, with lower levels of fat in muscle tissue had PCB and DDT contents lower than 0.1 ug/g [36].

As lipophilic molecules, organochlorines tend to be sequestered in lipid-rich tissues. Fat depots have higher organochlorine levels than other tissues [42]. The flesh of cod caught in Canadian Atlantic waters had mean concentrations of PCB and DDT of 0.04 ug/g and 0.024 ug/g, respectively. In the liver these levels were 5.1 ug/g and 5.2 ug/g [36]. Similarly, in cod caught at two sites in the same ocean area, the mean levels of DDT in the flesh were 0.049 ppm and 0.030 ppm, while the mean DDT contents of the livers were 8.1 ppm and 3.7 ppm [37].

Organochlorine Pesticides and Polychlorinated Biphenyls in Fish Oils

In comparison to their levels in fish, the concentration of PCBs and DDT in fish oils are significantly lower. In the oil prepared from cod livers with mean PCB and DDT concentrations of 5.1 ug/g and 5.2 ug/g, respectively, these compounds were found at levels of 2.5 ug/g and 1.2 ug/g [36]. In another comparison of the residue levels in cod liver and cod liver oil, the mean levels found in the oil represented only about one-tenth of the quantity suggested by the liver content [37]. In general, processed oils contain lower residue levels than would be suggested by their content in the raw materials [30,36,37].

The Removal of Organochlorine Compounds and Polychlorinated Biphenyls from Fish Oils

The processing of marine oils for edible use involves several steps: degumming by a phosphoric acid wash; removal of free fatty acids by alkali refining; bleaching; hydrogenation to an appropriate iodine value (and additional bleaching); and deodorization by stream stripping [30]. Organochlorine residues generally are not significantly reduced by degumming or alkali refining. The chemical degradation of DDT and DDD during hydrogenation reduces their content in oils. DDE and the PCBs which remain after hydrogenation are removed by steam distillation during deodorization [30].

A herring oil with initial levels of DDT and PCBs of 1.9 ppm and 9 ppm, respectively, showed similar contents after degumming, alkali refining, and bleaching. During hydrogenation, DDD and DDT were reduced by more than 90%, and their content fell to below detection levels. DDE and PCBs were reduced to 20% and 30% of their initial values, and deodorization removed the final trace amounts [39].

Similarly, in samples of crude redfish, flatfish, herring, mackerel, and dogfish oils obtained from commercial producers in Canada, DDT and its metabolites were removed to below detection levels during the processing of the oils to margarine stock. The initial content of DDT varied between 2 ppm and 8 ppm; PCB levels were between 3 ppm and 13 ppm. Neither DDT nor PCB residues were detected in the processed product; the detection level for compounds of the DDT group was 0.02 ppm and that for PCBs was 1 ppm [40]. It is evident that processing can effectively remove all detectable traces of these chlorinated hydrocarbons.

Deodorization by molecular distillation at 170°C removed 65% of the DDT residues in cod liver oil [40]. Thus even without hydrogenation, deodorization removes most of the DDT contaminant.

Conclusion

Organochlorine pesticides and polychlorinated biphenyls accumulate preferentially in fatty fish species and in the lipid-rich tissues of all fish. The processing of fish oils for edible use reduces the levels of these contaminants to make oils acceptable for human consumption. The use of accurate and rapid analytical methods and regular monitoring of processed fish oils can ensure consumer safety.

HEAVY METALS: LEAD, MERCURY, AND CADMIUM

The concentrations of lead, mercury, and cadmium have increased greatly in aquatic environments. Excessive intakes of lead and mercury can damage the central and peripheral nervous systems and kidney function [36]. Chronic exposure to cadmium causes emphysema, anemia, kidney disfunctions, and spontaneous fractures. In addition, epidemiological investigations have implicated cadmium as a carcinogen [35]. Both cadmium and lead have been associated with deaths due to heart disease [41].

Lead, Mercury, and Cadmium in Aquatic Organisms

In an aquatic environment, metals may exist as simple or complex ions in solution, associated with inorganic or organic particulates of varying size, and as chelates or colloids [32].

Most trace metals are taken up by aquatic organisms in the ionic form, which is the case with cadmium. The concentration of free cadmium ions appears to be the most important determinant of the extent of its uptake by fish [42]. Mercury and lead exist in both organic and inorganic forms. Their organic form is usually the methylated molecule synthesized by microorganisms [32]. The lipophilic nature of methylmercury and tetramethyl lead allows their accumulation by fish. The uptake of methylmercury from food and water, which resembles that of the organochlorines, occurs easily, and it subsequently is transported to lipid-rich tissues [32]. At higher trophic levels, greater concentrations of methylmercury occur in the tissues of fish [42].

Lead, Mercury, and Cadmium in Fish Oils and their Removal

The metal content of fish oil is influenced by the metal content of the fish tissue used as raw material and the production and storage conditions of the oil. The phospholipid content of the oil increases with longer storage [44], and during production, these compounds carry metals into the oils [43,45]. Thus, as the storage time of the raw fish prior to oil extraction is lengthened, the content of heavy metals in the extracted oil increases [43].

Although few studies of the metal content of fish oils before and after processing are available, processing oils for edible purposes appears to reduce the metal content [45,46]. The degree of reduction is dependent on the metal present. The removal of the bulk of a metal is usually accomplished in one refining or processing step [45]. Metals that are complexed to phospholipids are removed most easily by degumming and alkali refining since these steps remove phospholipids [45]. Further processing steps are less likely to remove metals that are tightly bound to oil constituents and organometallic compounds [46]. The FAO/WHO standards of acceptable metal content in oils have not always been met after processing [45].

Conclusion

The metal content of fish oils depends on the contamination of the source fish and the conditions of production and storage.

Few reports of the effect of processing on the metal content of fish oils are available. Although refining reduces the level of trace metals, the FAO/WHO standards may not be met. Further studies on the effect of processing on fish oils from diverse sources are needed.

REFERENCES

1. Lundberg, W.O. (1967). General deterioration reactions. In Fish Oils: their Chemistry, Technology, Stability, Nutritional Properties and Uses. M.E. Stansby (ed.) p. 141-147. Avi Publishing Co., Westport, Conn.

2. Olcott, H.S. (1967). Antioxidants. In Fish Oils: their Chemistry, Technology, Stability, Nutritional Properties and Uses. M.E. Stansby (ed.) p. 164-170. Avi Publishing Co., Westport, Conn.

3. Hardy, R. (1980). Fish lipids, Part 2. In Advances in Fish Science and Technology. J.J. Connell (ed.) p. 103-111. Fishing News Books, Farmham, Surrey, England.

4. Frankel, E.N. (1984). Lipid oxidation: mechanisms, products and biological significance. J. Am. Oil Chem. Soc. 61(12):1908-1917.

5. Schaich, K.M. (1980). Free radical initiation in proteins and amino acids by ionizing and ultraviolet radiations and lipid oxidation-Part III: free radical transfer from oxidizing lipids. CRC Crit. Rev. Food Sci. Nutr. 13:189-243.

6. Khayat, A. and D. Schwall. (1983). Lipid oxidation in seafood. Food Technology 37:130-140.

7. Gray, J.I. (1978). Measurement of lipid oxidation: a review. J. Am. Oil Chem. Soc. 55:539-546.

8. Logani, M.K. and R.E. Davies (1980). Lipid oxidation: biological effects and antioxidants-a review. Lipids 15(6):485-495.

9. Gryglewski, R.J. and S. Moncada (1979). Polyunsaturated fatty acids and thrombosis. Europ. J. Clin. Invest. 9:1-2.

10. Liston, J., M.E. Stansby, and H.S. Olcott (1963). Bacteriological and Chemical Basis for Deteriorative Changes. In Industrial Fishery Technology, p. 360-261. Reinhold Publishing Corp., New York.

11. Labuza, T.P. (1971). Kinetics of lipid oxidation in foods. CRC Critical Reviews in Food Technology 2(3):355-405.

12. Matsuo, N. (1954). Studies on the toxicity of fish oil. J. Biochem. (Tokyo) 41(4): 481-487.

13. Stansby, M.E. (1967). Odors and flavors. In Fish Oils: Their Chemistry, Technology, Stability, Nutritional Properties and Uses. M.E. Stansby (ed.) p. 171-180. Avi Publishing Co., Westport, Conn.

14. Lang, K. (1965). Biological properties of fish oils. In The Technology of Fish Utilization. R. Kreuzer (ed.) p. 223-224. Fishing News Books, London.

15. Kaneda T. and S. Ishii (1954). Nutritive value or toxicity of highly unsaturated fatty acids. J. Biochem. (Tokyo) 41(3):327-336.

16. Stansby, M.E. (1982). Properties of fish oils and their application to handling of fish and to nutritional and industrial use. In Chemistry and Biochemistry of Marine Food Products. R.E. Martin, G.J. Flick, C.E. Hebard, D.R. Ward (eds.) p. 75-92. Avi Publishing Co., Westport, Conn.

17. Bieri, J.G. (1984). Sources and consumption of antioxidants in the diet. J. Am. Oil Chem. Soc. 61(12):1917-1918.

18. Witting, L.A. (1970). In Progress in the Chemistry of Fats and other Lipids. R.T.Holman (ed.) Pergamon Press, Oxford, U.K.

19. Christopherson, B.O., J. Norseth, M.S. Thomassen, E.N. Christianson, K.R. Norum, H. Osmundsen, J. Bremer (1982). Metabolism and metabolic effects of C22:1 fatty acids with special reference to cardiac lipidosis. In Nutritional Evaluation of Long Chain Fatty Acids in Fish Oil. S.M. Barlow and M.E. Stansby (eds.) p. 89-139. Academic Press, London.

20. Svaar, H. (1982). The long-term heart lesion phenomenon in animals and humans. In <u>Nutritional Evaluation of Long-Chain Fatty Acids in Fish Oil</u>. S.M. Barlow and M.E. Stansby (eds.) p.163-184. Academic Press, London.

21. Barer, R. (1982). Lipidosis in the rodent heart and other organs. In <u>Nutritional Evaluation of Long-Chain Fatty Acids in Fish Oil</u>. S.M. Barlow and M.E. Stansby (eds.) p. 141-161. Academic Press, London.

22. Duthie, I.F. and S.M. Barlow (1982). A rat life span study comparing partially hydrogenated fish oils, partially hydrogenated soybean oil and rapeseed oil included in the diet at high levels: outline description and interim communication. In <u>Nutritional Evaluation of Long-Chain Fatty Acids in Fish Oil</u>. S.M. Barlow and M.E. Stansby (eds.) p. 185-214. Academic Press, London.

23. Ackman, R.G. (1980). Fish lipids Part 1. In <u>Advances in Fish Science and Technology</u>. J.J. Connell (ed.) p. 86-102. Fishing News Books, Farmham, Surrey, England.

24. Ackman, R.G. (1982). Fatty acid composition of fish lipids. In <u>Nutritional Evaluation of Long-Chain Fatty Acids in Fish Oil</u>. S.M. Barlow and M.E. Stansby (eds.) p. 25-88. Academic Press, London.

25. Barlow, S.M., and I.F. Duthie (1984). An evaluation of the safety of partially hydrogenated marine oils in the human diet. <u>Nutrition Abstracts and Reviews</u> 54(1):17-30.

26. Norum, K. (1982). Discussion and summing up. In <u>Nutritional Evaluation of Long-Chain Fatty Acids in Fish</u>. S.M. Barlow and M.E. Stansby (eds.) p. 283-314. Academic Press, London.

27. Duthie, I.F. (1982). Discussion and summing up. In <u>Nutritional Evaluation of Long-Chain Fatty Acids in Fish</u>. S.M. Barlow and M.E. Stansby (eds.) p. 282-314. Academic Press, London.

28. Carpenter, K.J. (1980). Fish in human and animal nutrition. In <u>Advances in Fish Science and Technology</u>. J.J. Connell (ed.) p.86-102. Fishing Book News, Farmham, Surrey, England.

29. Goodnight, S.H., W.S. Harris, W.E. Connor and D.R. Illingworth (1982). Polyunsaturated fatty acids, hyperlipidemia, and thrombosis. <u>Arteriosclerosis</u> 2(2):87-113.

30. Addison, R.F. (1982). Organochlorine compounds and marine lipids. <u>Prog. Lipid Res.</u> 21:47-71.

31. Johnson, K. (1982). Equity in Hazard Management. <u>Environment</u> 24(9):28-38.

32. Phillips, D.J.H. (1980). Quantitative Aquatic Biological Indicators. Applied Science Publishers Ltd., London.

33. Sun, Marjorie. (1983). EPA, Utilities Grapple with PCB Problems. Science 222:32-33.

34. Sprague, J.B. and J.R. Duffy (1971). DDT Residues in Canadian Atlantic fishes and shellfishes in 1967. J. Fish. Res. Bd. Canada 28:59-64.

35. Koller, L.D. (1980). Public health risks of environmental contaminants: heavy metals and industrial chemicals. J. Am.Vet. Med. Assoc. 176(6): 525-529.

36. Sims, G.G., J.R. Campbell, F. Zemlyak, and J.M. Graham (1977). Organochlorine residues in fish and fishery products from the Northwest Atlantic. Bull. Environ. Contam. & Toxic. 18:697-705.

37. Sims, G.G., C.E. Cosham, J. R. Campbell, and M.C. Murray (1975). DDT residues in cod livers from the Maritime Provinces of Canada. Bull. Environ. Contam. & Toxic. 14:505-512.

38. Julshamn, K., L. Karlsen, and O.R. Braekkan (1973). Removal of DDT and its metabolites from fish by molecular distillation. Fiskeridirektoratets Skrifter Serie Teknologiske Undersokelser 5(15):1-12.

39. Addison, R.F. and R.G. Ackman (1977). Stepwise removal of chlorinated hydrocarbons during processing of herring oil for edible use. J. Am. Oil Chem. Soc. 54:153A.

40. Addison, R.F. and R.G. Ackman (1974). Removal of organochlorine pesticides and polychlorinated biphenyls from marine oils during refining and hydrogenation for edible use. J. Am. Oil Chem. Soc. 51: 192-194.

41. Voors, A.W., W.D. Johnson, and M.S. Shuman (1982). Additive statisitical effects of cadmium and lead on heart-related disease in a North Carolina autopsy series. Arch. Environ. Health 37:98-102.

42. Luoma, S.N. (1983). Bioavailability of trace metals to the aquatic environment- a review. Sci. Tot. Environ. 28:1-22.

43. Lunde, G., L.H. Landmark, J. Gether (1976). Sequestering and exchange of metal ions in edible oils containing phospholipids. J. Am. Oil Chem. Soc. 53:207-210.

44. Lunde, G. (1973). The analysis of organically bound elements and phosphorus in raw, refined, bleached and hydrogenated marine oils produced from fish of different quality. J. Am. Oil. Chem. Soc. 50:26-28.

45. Elson, C.M. and R.G. Ackman (1978). Trace metal content of a herring oil at various stages of pilot plant refining and partial hydrogenation. J. Am. Oil Chem. Soc. 55:616-618.

46. Elson, C.M., E.M. Bem. and R.G. Ackman (1981). Determination of heavy metals in a menhaden oil after refining and hydrogenation using several analytical methods. J. Am. Oil Chem. Soc. 58: 1024-1026.

47. Kanner, J., B. German, and J.E. Kinsella (1986). Initiation of lipid peroxidation in biological systems: a review. CRC Reviews in Food Science and Nutrition. (In review.)

9
Edible Fish Oil Processing and Technology

INTRODUCTION

If fish oils do exert beneficial effects as discussed in the preceding sections and if a dose of 5g per day is generally recommended, then a market for approximately 425 million kg of edible fish oil could develop in the United States alone. This potential market raises questions concerning the future sources and supply of fish oil.

Currently fish oil production in the United States is a byproduct of the fish meal industry. Fish oil is produced from offals (fish remains after filleting) and from industrial fish, such as sardine, mackerel, tuna, and menhaden. The oil content of these species is listed in Table 9.1 and their fatty acid content in Table 9.2.

CATCH

In 1984 of the 6.4 billion pounds of edible and industrial fish caught in U.S. ports, 2.9 billion pounds were menhaden, which represent 45% of the commercial fishery landings [1]. Menhaden is used mostly in oil and meal production [2], and Cameron, Louisiana, and Pascagoula-Moss Point, Mississippi, are the leading ports in quantity of commercial fishery landings [1].

TABLE 9.1 Oil Content of Principal Fish Used in Fish Oil Production.

		%	Average
Salmon	pink	1.8 - 12.5	4.8
	red	1.6 - 13.2	6.9
	silver	1.3 - 9.9	5.7
Mackerel	Pacific	0.3 - 15.9	6.6
	Atlantic	0.7 - 24.0	9.6
	Spanish	0.6 - 14.4	5.9
Menhanden	Atlantic	10.2	10.2
	gulf	2.1 - 17.8	11.8
Tuna	Albacore	0.7 - 13.2	6.3
	bluefin	0.5 - 14.1	5.0
	skipjack	0.2 - 11.0	4.5

Source: Chapter 11

PRODUCTION

In 1984, 372.7 million pounds of fish oil valued at $61 million were produced. Menhaden oil accounted for 98% of this volume [1]. From 1983 to 1984, tuna and mackerel production decreased by 867,000 pounds, and unclassified oil, including anchovy, decreased by 5.9 million pounds (Table 9.3). Most of the fish oil exported from the United States in 1983 and 1984 was sent to the Netherlands and England where it was processed mostly for margarine (Table 9.4).

During 1980-83, the United States ranked third in production of fish oils behind Japan and Norway [2]. In 1983, the United States produced 399.3 million pounds of fish oil, a record compared to other years between 1975 and 1984. In 1982 the United States had a very high utilization of fish oil (Table 9.5), mostly as an ingredient in soaps, paints and varnishes, floor coverings, oilcloth, printing inks, rubber, lubricants, leather, insecticides, and cosmetics [3].

TABLE 9.2 Fatty Acid Composition of Principal Fish Used in Fish Oil Production

Fatty Acid (wt %)	Salmon		Fish		Tuna	
	Pink	Silver	Mackerel	Menhaden	Albacore	Bluefin
14:0	3.4	3.7	4.9	8.0	3.7	4.5
15:0	1.0	NK	0.5	0.5	1.0	0.6
16:0	10.2	22.5	28.2	28.9	29.3	22.1
16:1	5.0	5.0	5.3	7.9	6.3	2.8
17:0	1.6	NK	1.0	1.0	1.2	0.8
18:0	4.4	3.9	3.9	4.0	6.1	6.2
18:1	17.6	23.6	19.3	13.4	16.6	21.7
18:2	1.6	0.9	1.1	1.1	0.7	0.8
18:3	1.1	1.0	1.3	0.9	0.6	NK
18:4	2.9	NK	3.4	1.9	2.2	0.9
20:1	4.0	5.3	3.1	0.9	2.7	6.3
20:4	0.7	1.4	3.9	1.2	1.2	1.0
20:5	13.5	11.0	7.1	10.2	6.5	6.4
22:1	3.5	NK	2.8	1.7	2.0	5.4
22:5	3.1	3.3	1.2	1.6	0.8	1.4
22:6	18.9	12.9	10.8	12.8	17.6	17.1

NK = not known

Source: Ref. 39

TABLE 9.3 Production of Fish Meal, Oil and Solubles (1983, 1984).

	1983		1984	
Product	Thousand pounds	Thousand dollars	Thousand pounds	Thousand dollars
Body Oil				
Menhaden (1)	385,779	64,318	365,895	60,011
Tuna and mackerel	2,535	557	1,668	209
Unclassified (2)	11,020	1,939	5,155	731
Total	399,334	66,814	372,718	60,951

(1) May include small quantities made from other species
(2) Includes anchovy

Source: Ref. 1

TABLE 9.4 Domestic Fish and Marine Oil Exports, by Country of Destination, 1983 and 1984.

	1983		1984	
Country	Thousand pounds	Thousand dollars	Thousand pounds	Thousand dollars
Netherlands	172,366	23,647	277,552	50,254
United Kingdom	117,246	16,866	32,322	6,393
Belgium, Luxembourg	18,741	2,074	29,154	4,648
Republic of S. Africa	5,730	835	27,315	4,381
Sweden	7,784	1,044	16,966	2,352
Fed. Republic of Germany	27,707	3,548	9,773	1,270
Columbia	8,821	1,531	4,423	935
Canada	706	280	590	275
Spain	(1)	3	1,101	245
Other	44,986	10,008	229	228
Total	404,087	59,836	399,425	70,981

(1) Less than 500 lb.

Source: Ref. 1

TABLE 9.5   U.S. Supply of Fish Oils, 1975-1984.

Year	Domestic production	Imports (1)	Total supply	Exports	Total U.S. consumption
	------------------- Thousand pounds ---------------------
1975	245,653	11,283	256,936	191,843	65,093
1976	204,581	20,937	225,518	179,235	46,283
1977	133,182	13,731	146,913	90,633	56,280
1978	296,287	16,040	312,327	222,012	90,315
1979	267,949	14,455	282,404	198,497	83,907
1980	312,511	21,350	333,861	284,009	49,852
1981	184,302	18,255	202,557	238,308	(2)
1982	347,513	12,699	360,212	202,345	157,867
1983	*399,334	15,334	414,668	404,087	10,581
1984	372,718	13,426	386,144	399,425	(2)

(1) Excludes fish liver oils
(2) The 1981 and 1984 exports, which included prior year stocks, exceeded domestic production plus imports.
* Record.

Source: Ref. 1

PROCESSING

The edible fish oil industry in the United States began in Rhode Island in 1811. By 1855 many of the current processing methods were being used and by 1911 had been adopted nationwide. Since 1911 the techniques used for oil recovery and refining have changed little, but improvements in speed of processing, uniformity of quality, minimization of rancidity, reduction of off-flavors, and overall improvement in shelf life have been made [3].

The process and processing equipment used depends upon the condition, species, and size of the fish, its oil content, the season, and the temperature [4]. These factors influence speed of extraction as well as cooking time and temperature [5].

In processing fish for edible use, the first and most important requirement is raw material of superior quality. The fish should be undamaged and chilled from the time of catch to prevent deterioration.

To begin processing, fish are minced and transferred to a cooker via a feed screw. Cooking causes protein denaturation, breakdown of tissue, and release of oil and is the most crucial step in processing. Undercooking results in reduced production and overcooking causes excessive breakdown of muscle tissue. Typical cooking conditions are 15 minutes in the cooker at 90°C [5].

From the cooker, the released oil is transferred via a strainer to a screwpress filter. The liquor produced, which is composed of oil and stickwater, is then passed through a screen, and the liquor is separated at 90°C by desludging in self-cleaning centrifuges. Next, a hot water (90-95°C) wash (10% of oil volume) is used to remove trace amounts of protein materials. At this point, the crude oil contains moisture and volatiles (0.11%), insoluble impurities (0.01%), and free fatty acids (0.45%) [6].

The quality of crude fish oil is dependent on the storage and handling of the fish prior to processing, the quality and efficiency of the processing plant, and its storage and handling after processing.

Impurities present in crude oil are classified as soluble, insoluble, or colloidal. Soluble impurities include metals, pigments, chemicals, and free fatty acids; insoluble impurities are moisture, dirt, and rust; colloidal impurities are mostly proteins [5]. Quality control tests are done at this point to ensure that the crude fish oil matches contract and refiner's specifications.

The crude fish oil next is processed for either edible or industrial uses. Figure 9.1 is a schematic representation of the refining of crude oil for edible use.

Refining begins in the storage tank where the oil is agitated slowly at about 25°C to provide an homogenous product. Degumming, in which phosphatides are insolublized with 80% phosphoric acid, is then performed [5]. The phosphorous levels must be reduced to less than 1

Physical Refining   Crude Oil
Degumming
Neutralization - alkali refining
Water washing
Drying
Bleaching
Filtration
Pretreated or neutralized
 and bleached storage
Hydrogenation (if desirable)
Deodorization
Polishing
Cooling
Edible Oil Storage

FIGURE 9.1 Sequence of steps involved in fish oil refining (from Ref. 5).

ppm [7]. The sludge formed is then removed by centrifugation, and the oil is neutralized with caustic soda. The use of a self-cleaning centrifuge in the production of the crude oil usually eliminates this step.

There are two methods of refining. The conventional caustic soda method--alkali refining, or neutralization--is carried out on a centrifugal continuous refining line. The oil is treated with sodium hydroxide (6.6% w/w [8]; 1N to 6N [9]; 4N [5], 7% to 18% [10], 15% to 18% NaOH [11]), which neutralizes the free fatty acid content to about 0.01% to 0.03% [9] and reacts with impurities to make them soluble and therefore removable in the aqueous phase. After centrifugation, this step may be repeated if the oil is of inferior quality. The oil is next washed with water to remove the soaps and centrifuged. Sometimes phosphoric or citric acids are added to the water wash to ensure the removal of metal ions and soaps that interfere with subsequent bleaching and hydrogenation [5]. The oil is then dried to remove water.

In the physical refining method, a distillation step at 250°C under 2mm to 5 mm absolute pressure is employed after degumming to remove fatty acids, heat-degradable pigments, and other impurities.

TABLE 9.6 Refining of Oils - Process and Impurities Removed.

Stage	Impurities reduced or removed
Crude oil storage	oil insoluble materials
Degumming	phospholipids, sugars, resins, proteinaceous compounds, trace metals and others
Neutralization	fatty acids, pigments, phospholipids, sulphur compounds, oil insoluble materials, water soluble materials
Washing	soaps
Drying	water
Bleaching	pigments, oxidation products, trace metals, trace soaps
Filtration	spent bleaching earth
Deodorization	fatty acids, mono- and diglycerides, aldehydes and ketones, hydrocarbons, sulphur compounds, pigment decomposition products
Physical refining	fatty acids, mono- and diglycerides, aldehydes and ketones, hydrocarbons, sulphur compounds, pigment decomposition products
Polishing	removal of trace oil insoluble materials (to 30 um or 10 um)

Source: Ref. 5

Although this method produces higher yields, it has not yet proved practical due to variability in quality and amount of impurities [7], especially sulphur, in the fish oil.

After refining, the oil is bleached with earth, clay, or charcoal (natural clay, activated clay, and activated carbon) to remove pigments and reduce oxidation products, soaps [9], phospho-

Edible Fish Oil Processing and Technology 217

rous, odorous substances [12], and trace metals. Bleaching clays also improve hydrogenation [13]. Semicontinuous and continuous plants operate under vaccuum at 90° to 110°C with the amount of clay varying between 0.2% to 3.0%, depending on oil quality. The bleaching earths contain Mg, Ca, Fe, and Na, and they are activated with mineral acids that increase bleaching activity and oil retention [14]. Removal of the clay by filtration is essential.

After removal of the clay, the oil is deodorized by steam distillation under vacuum. This step removes disagreeable flavors, odors, and volatile constituents by steam stripping the oil of free fatty acid components, ketones, and aldehydes to approximately 0.1% concentration [12,15]. As the system pressure and steam rate are lowered, free fatty acid reduction is improved [15]. The oil is released at 60°C and pumped via a polishing filter and cooler to storage. Table 9.6 lists the impurities removed at each step of processing.

The oil should be freshly deodorized, bland in odor and flavor, and have a free fatty acid content of less than 0.03% [15], a peroxide value of 0, and a color of less than 2.0 red and 20 yellow units in a Lovibond cell [5]. The fatty acid composition of commercial menhaden oil is given in Table 9.7.

For use as a nutritional supplement or in therapeutic applications, the oil has to be carefully refined and its vitamin A and D content reduced, supplemented with antioxidants, and then packaged to minimize autoxidation.

For shortening/margarine uses the oil is hydrogenated before deodorization. At this point the product is checked again for free fatty acids (0.12% max); soap (0); phosphorous (4 ppm max); sulphur (15 ppm max); and Totox value (12 max: Totox = 2x peroxide value + anisidine value) [5].

Hydrogenation is carried out with nickel catalysts in a batch plant. Hydrogen is compressed to 3 atmospheres and fed to the converter containing the oil and catalyst. The temperature is 150/160°C for the first hour and then maintained at 180/190°C. The reaction is controlled by intermittent sample analysis for melting

TABLE 9.7 Variation in Fatty Acids of Commercial Menhaden Oil

Fatty Acid (wt %)	Minimum	Maximum	Average
14:0	7	16	10
15:0 and 17:0	2	4	3
16:0	19	24	21
16:1	11	18	12
18:0	2	3	3
18:1	10	23	12
18:2	2	3	3
18:3	0.4	1.7	1
18:4	0.8	3.6	3
20:1	1.6	2.7	2
20:4	0.6	2.1	2
20:5	10	15	13
22:5	0.3	2.5	2
22:6	3.3	11	8

Source: Ref. 40

point, iodine value, and refractive index. Upon completion of this step, the oil is cooled to 90°C, treated with clay and citric acid to remove the nickel, and filtered, and the recovered catalyst is returned for reuse [14].

Edible Fish Oil Processing and Technology 219

PROBLEMS INVOLVED IN PROCESSING

The problems involved in processing fish oils fall into two categories: removing impurities from the oil and stabilizing the oil to prevent oxidation and off-flavor production.

Removal of impurities involves removal of trace metals, minerals, and pesticides. Trace metals and minerals are removed by degumming, alkali refining [17], and by bleaching [7]. Selenium is significantly lowered by hydrogenation and deodorization. Alkali refining removes arsenic and reduces lead by 40%. Usually the bulk of a trace element is removed in one processing step [18]. With successive refining steps, the metal content is reduced to a level which meets FAO/WHO codex standards [8]. Sulphur levels are reduced 25% during neutralization and 25% during bleaching [7].

Quantitative removal of DDT, DDE, aldrin, and dieldrin has been reported [12]. Initial levels in crude oil (2ppm to 8 ppm DDT; 0 to 0.003 ppm dieldrin; and 3 ppm to 13 ppm PCBs) were reduced below detectable limits as a result of processing [19], and deodorization removed those pollutants not removed by hydrogenation [20].

Stability of fish oil is the major factor that determines quality. Autoxidation (see chapter 8) of the unsaturated fatty acid components forms highly unsaturated hydroperoxides that are extremely unstable and break down to yield off-flavor components [21,22]. For this reason, fish oil processing must produce a very pure product with very low hydroperoxide levels.

Conditions which enhance oxidation are high temperature, light, oxygen exposure, certain metals [23], and heme pigments [24]. Initially, if the oil is refrigerated, protected from light, and stored in inert, airtight containers, oxidation should be very slow. However, to minimize oxidation, antioxidants are used [25].

Antioxidants are compounds which delay the oxidation process by interfering either with the intitiation steps of oxidation (induction period) or with the early propagation steps [26]. They react with the free radical formed in the initial or early steps to give an intermediate that terminates the process. Antioxidants such as tocopherols occur naturally in fish oils in small amounts but are lost in

TABLE 9.8 Characteristics of Various Antioxidants

Characteristics	Antioxidant	
	Butylated hydroxy anisole (BHA)	Butylated hydroxy toluene (BHT)
Form	synthetic [1] waxy powder [2]	powder [2]
Chemistry	mpt 48-65°C; insoluble in water, 25% soluble in ethanol; 100% soluble in propylglycol; 25-40% soluble in veg. oil [2]; prolongs inductive and oxidative phase [4]; legal in U.S., used in conc.<0.02%	insoluble in water or water or propylglycol; 25% soluble in ethanol; legal in U.S., used in conc.<0.02% [3]
Synergist	ascorbic acid	ascorbyl palmitate [3]
	Gum Guaiac	Propyl Gallate
Form		synthetic [1] crystalline powder [2]
Chemistry	legal in U.S., used in conc.<0.02% [2]	slightly soluble in water; 30% soluble in propylglycol [2]; legal in U.S., conc.<0.02% [3]
Synergist	lecithin and phosphoric acid [3]	citric acid [3]
	Nordihydroguaretic Acid	Tocopherol (Vitamin E)
Form		natural viscous oil [2]
Chemistry	legal in U.S., conc. <0.02% [3]	not soluble in water; miscible in ethanol and veg. oils [2]; effective at various humidities [5]; more effective than sterilizing or refining [6]
Synergist	isopropyl citrate [3]	

processing. The tocopherol content of various fish oils ranges from 40 mg tocopherol/kg fish in sardine oil and 66 mg tocopherol/kg fish

Table 9.8 (continued)

Characteristics	Antioxidant	
	Ascorbyl Palmitate	Tertbutylhydroquinone
Form	powder [2]	
Chemistry	very slightly soluble in water; 22% soluble in ethanol; 10% in propylglycol emulsifiers, 400 ppm in veg. oils [2]	legal in U.S. in 1972 [7]

1. Brody, J. (1965). Fishery By-Products Technology. Avi Publishing Co., Inc., Westport, Conn. pp. 147-164.
2. Gregory, D. (1984). The radical answer. Food Flav. Ingred. Proc. 6:18-24.
3. Olcott, H.S. (1967). Antioxidants. in Fish Oils. pp.164-70. M.E. Stansby, ed. Avi Publishing Co., Inc. Westport, Conn.
4. Sen, D.P. and R.A. Padival (1970). Effect of certain proteins on the stabilization of fish oil. J. Food Sci. Tech. 7:153-58.
5. Takahashi, K., et al. (1983). Antioxidantive effect of tocopherol isomers against the oxidation of fish lipid under various water activity conditions. Bull. Fac. Fish. Hokkaido Univ. 34(2):124-30.
6. Gheyasuddin, S., M.A. Rahman, and A.K.M.A. Bhuiyan (1978). Chemical characteristic of hilsha fish oil and keeping quality. Bangladesh J. Biol. Sci. 6 and 7(1):10-17.
7. White, P.A. (1984). The problem solvers, antioxidants. Food Flav. Ingred. Proc. 6:18-24.

in menhaden oil to 142 mg tocopherol/kg fish in herring oil and 217 mg tocopherol/kg fish in pink salmon oil [27] (Chapter 7).

Added antioxidants should be highly effective at low concentrations, reasonably priced, and readily soluble; they should be approved for use in food and should not mask the effect of improper processing or storage or contribute to off flavors [25]. Antioxidants are usually added directly at approximately 0.02% by weight (Table 9.8). Some antioxidants used under experimental conditions in sardine oil are listed in Table 9.9.

TABLE 9.9 Antioxidants Used in Sardine Oil to Improve Stability.

Antioxidant	Characteristics	Reference
L-proline	0.02% level as effective as BHA	1
	0.1% level more effective than BHA	2
Spice mixture oleoresin (cinnamon, red chillies tumeric, ginger, black pepper, clove, cumin and mustard seeds)	5g spice/100 ml ethylalcohol, heated washed and distilled 0.5%, 1.0% better antioxidant than 0.02% BHA and 0.001% citric acid	3
20% hydroxytoluene, 20% hydroxyanisole, 10% monoglyceride citrate added at 0.05% level	with nitrogen bubbling, after 7 months, still stable	4

1. Revankar, G.D. (1974). Proline as an antioxidant in fish oil. J. Food Sct. Tech. 11:10-11.
2. Noguchi, S. and Matsumoto (1975). Studies on the control of denaturation of fish muscle proteins during frozen storage. Bull. Jap. Sci. Fish. 41(2):243-49.
3. Revankar, G.D. and D.P. Sen (1974). Antioxidant effect of spice mixture on sardine oil. J. Food Sci. Tech. 11:31-32.
4. Hung, S.S.O. and S.J. Slinger (1981). Studies of chemical methods for assessing oxidative quality and storage stability of feeding oils. J. Am. Oil Chem. Soc. 58:785-88.

ADVANCES IN PROCESSING OF FISH OILS

It is often desirable to produce fish oils more highly enriched with PUFA fractions. High EPA concentrations are desirable in fish oils used for clinical therapy or where high intake of fish oils is not tolerated. Simple fractionation processes do not discriminate among the PUFAs and are not useful for enriching oils in EPA. To prepare fractions high in EPA, several methods have been developed. Due to high cost, however, these are most suitable for

Edible Fish Oil Processing and Technology

small-scale, laboratory preparation of purified samples rather than to the processing of large amounts of fish oil. Techniques amenable to small-scale preparation include urea adduct fractionation [36] and silver resin chromatography [37]. Other chromatographic techniques require large volumes of solvents that add to the cost of fractionation.

A new method recently used to process fish oil on a laboratory scale is supercritical gaseous (carbon dioxide) extraction and fractionation. Supercritical CO_2 extraction as described by Eisenbach [28] Nelson [29], and Rizvi [38] can be useful in producing pure fish oils and in deodorizing fish oils to yield a lighter colored, milder flavored oil. The CO_2 is liquefied by compression at pressures from 4 to 10,000 psi. This liquid is then passed through an extraction vessel containing the oil. Specific fractions can be recovered and isolated by varying pressure or temperature [28]. The fractions are highly concentrated, unsaturated oil components.

Fractionation is used in refining processes to remove particular components, such as those with higher melting points [30]. It also can be used to process fish oil for use in its natural liquid state, such as for salad oils. Ensuring stability for such use involves light hydrogenation to an iodine value of 105-110 and then fractionation to give a liquid oil of acceptable cold test and improved shelf life [7]. However, this reduces the 20:5 and 22:6 fatty acid components.

QUALITY CONTROL STANDARDS FOR FISH OILS

In 1979, the International Association of Fish Meal Manufacturers published a bulletin describing the standard tests used by refiners of fish oil destined for edible use.

The first set of tests used to assess the quality of the oil include the following criteria: moisture content (high moisture could lead to rust formation, oxidation, and color problems); dirt (visually inspected); appearance (golden brown is easy to refine, dull brown causes problems in refining, and frothiness causes emulsification problems); free fatty acid (by titration, (AOAC methods [31]); and iodine value (AOAC methods [31]).

TABLE 9.10 Refiner's Quality Assurance Criteria

% Moisture		0.3% maximum
% Suspended solids		0.1% maximum
% FFA		4.0% maximum
Peroxide value (m. equivs./kg)		6.0 maximum
Anisidine value		12.0 maximum
U.V. Extinction $E_{1cm}^{1\%}$ 233 nm		10.0 maximum
$E_{1cm}^{1\%}$ 269 nm		4.0 maximum
Iron content		1.5 ppm max
Cooper content		0.2 ppm max
Sulphur	crude oil	30.0 ppm max
	neutralized oil	15.0 ppm max
Phosphorous		0.02% maximum

Source: Ref. 7

The second set of tests, which aids in the refining process, determines peroxide value, anisidine value, and thiobarbituric acid test (TBA) values to establish concentrations of oxidation products; U.V. extinction, as a guide to bleachability and indication of overheating [5]; and the presence of trace metals, iron, and copper, which are catalysts for oxidation. These tests give the refiner certain quality control standards that are used for reference (Table 9.10).

STORAGE AND PACKAGING OF FISH OILS

Fish oil is stored in dark, noncorrodable steel or preferably aluminum tanks [32] under nitrogen or carbon dioxide to exclude oxygen at a low constant temperature. The tank is drained from the bottom to separate residual water. In cold climates, steam heated coils are used to keep the oil fluid. To ensure good settling, the tanks are preferably insulated, and to prevent contamination, the tanks are unloaded from a tube situated somewhere above the bottom of the tank [32].

More research on packaging fish oil is needed, especially on improved packaging methods for individual dosage as well as for bulk

Edible Fish Oil Processing and Technology

storage. When packaging for individual dosage, the loss of nutritional value and stability, the introduction of oxygen and other impurities, and the exchange of materials between packaging and oil must be avoided. Flavor deterioration is a major problem [21].

MARKETING OF FISH OILS

Up to now fish oil has been marketed for its use in the production of margarine. In 1925, 120,000 pounds of California sardine oil was used to make margarine. This figure continued to increase until 1951 when federal regulations curtailed its use and sanitary requirements could not be maintained at fish processing plants [12]. Production ceased in the United States and increased in Canada, Japan, and most European countries [33].

Problems that need to be overcome in the marketing of fish oil include flavor reversion, quality, and public image. To the U.S. consumer, a fishy oil flavor means lower quality compared to products made from vegetable oils. In Scandinavia and Northern Europe, however, this flavor is quite acceptable [12]. Education and advertising is needed to change the image of fish oil in the United States. The marketing strategy for fish oils should include educating consumers about the prophylactic and therapeutic benefits of fish oils in the prevention and treatment of cardiovascular disease.

Once sanitation and preservation problems have been dealt with and the quality of the oil ensured, the question of affordability must be met. Presently fish oil is cheap, but with modifications to improve quality and the development of a market, the prices may increase.

INDUSTRIAL NEEDS

Physical Needs

Good storage facilities with sufficient refrigeration at the docks are needed to ensure maintenance of fish quality and to enable more than one day's supply of fish to be bought or sold. Storage in CO_2 atmosphere would increase market life [34].

Full service fishing ports would ensure high yield and rapid turnover by quick unloading, processing at the dock versus transporting, and preselling products versus auctioning upon arrival [35].

Fish oil processing plants must meet more stringent sanitation codes so that fish oil can be used for edible purposes in the United States.

Research Needs

The source and nature of impurities must be investigated and reduced in the storage and processing of fish oils. Corroded tanks, for example, and rusting cause oxidation, so aluminum tanks and possibly aluminum equipment should be used.

Research is needed to identify superior antioxidants and optimize concentrations of appropriate antioxidant mixtures required to maintain the shelflife of fish oil.

Because the most effective clinical dosages of n-3 PUFAs require high intake levels, it is desirable to enrich the active EPA component so dosage can be reduced. New fractionation techniques that are inexpensive and adaptable to large-volume fish oil processing need to be developed.

Better packaging materials and storage conditions must be developed for oil intended for human consumption, for example individual dosage or encapsulation. Strict quality control standards on the shelflife of these products also must be established.

Different fish need to be studied to determine the availability of the most appropriate fish for edible human oil (see chapter 11).

The public needs to be educated about the nutritional value of fish oils and their potential beneficial effects in reducing heart disease. The consumer's image of fish must improve in this country if fish oils are to become a viable industry. This public image depends on consistent quality.

REFERENCES

1. U.S. Department of Commerce, National Oceanic and Atmospheric Administration (1985). <u>Fisheries of the United States, 1984</u>. Current Fishery Statistics, No. 8360.

2. Food and Agricultural Organization of the United Nations (1983). <u>Yearbook of Fishery Statistics. Fishery Commodities</u>, volume 57.

Edible Fish Oil Processing and Technology

3. Stansby, M. E. (1978). Development of fish oil industry in the United States. J. Am. Oil Chem. Soc. 55: 238-43.

4. Aure, L. (1967). Manufacture of fish-liver oil. In Fish Oils, pp. 193-205. M. E. Stansby (ed), Avi Publishing Co., Inc., Westport, Conn.

5. Young, F. V. K. (1982). The production and use of fish oils. In Nutritional Evaluation of Long-Chain Fatty Acids, pp. 1-23. S. M. Barlow and M. E. Stansby (eds.), Academic Press, London.

6. Pigott, G. M. (1967). Production of fish oil. In Fish Oils, pp. 183-92. M. E. Stansby (ed.), Avi Publishing Co., Inc. Westport, Conn.

7. International Association of Fish Meal Manufacturers (1979). Fish Oil Bulletin, No. 10. Hoval House, Orchard Parade, Mutton Lane, England.

8. Elson, C. M., E. M. Bem, R. G. Ackman (1981). Determination of heavy metals in a menhaden oil after refining and hydrogenation using several analytical methods. J. Am. Oil Chem. Soc. 58: 1024-26.

9. Chang, S. S. (1967). Processing of fish oils. In Fish Oils, pp. 206-21. M. E. Stansby (ed.), Avi Publishing Co., Inc., Westport, Conn.

10. Brody, J. (1965). Fishery By-Products Technology. Avi Publishing Co., Inc., Westport, Conn. pp.179-187.

11. Klein, K. and L. S. Crauer (1974). Further developments in crude oil processing. J. Am. Oil Chem. Soc. 51: 382A-84A.

12. Gauglitz, E. J., Jr., V. F. Stout, and J. C. Wekell (1974). Application of fish oils in the food industry. In Fishery Products, pp. 132-36. R. Kreuzer (ed.), Fishing News Books, Surrey, England.

13. Zschau. W. (1984). The use of bleaching earth in fatty acid production. J. Am. Oil Chem. Soc. 61:214-18.

14. Gunstone, F. D. and F. A. Norris (1983). Lipids in Foods. Chemistry, Biochemistry and Technology. Pergamon Press, New York.

15. Gavin, A. (1977). Edible oil deodorizing systems. J. Am. Oil Chem. Soc. 54: 528-32.

16. Stansby, M. D. (1971). Flavors and odors of fish oils. J. Am. Oil Chem. Soc. 48: 820-23.

17. Notevarp, O. and M. H. Chahine (1972). Trace metal contents, chemical properties and oxidative stability of capelin and herring oils produced in Norwegian plants. J. Am. Oil Chem. Soc. 49: 274-77.

18. Elson, C. M. and R. G. Ackman (1978). Trace metal content of a herring oil at various stages of pilot plant refining and partial hydrogenation. J. Am. Oil Chem. Soc. 555: 616-18.

19. Addison, R. F. (1974). Removal of organochlorine pesticides and polychlorinated biphenyls from marine oils during refining and hydrogenation for edible use. J. Am. Oil Chem. Soc. 52: 192-94.

20. Addison, R. F., M. E. Zinck, R. G. Ackman, and J. C. Shipos (1978). Behavior of DDT, polychlorinated biphenyls (PCBs) and dieldrin at various stages of refining marine oils for edible use. J. Am. Oil Chem. Soc. 55: 391-94.

21. Connell, J. J. (ed.) (1980). Advances in Fish Science and Technology. Fishing News Books Ltd. Farnham, Surrey, England.

22. Hung, S. S. O. and S. J. Slinger (1981). Studies of chemical methods for assessing oxidative quality and storage stability of feeding oils. J. Am. Oil Chem. Soc. 58: 785-88.

23. Lundberg, W. O. (1967). General deterioration reactions. In Fish Oils, pp. 141-47. M. E. Stansby (ed.), Avi Publishing Co., Inc., Westport, Conn.

24. Zama, K., K. Takama, and Y. Mizushima (1979). Effect of metal salts and antioxidants on the oxidation of fish lipids during storage under the conditions of low and intermediate moistures. J. Food Preserv. 3: 249-57.

25. Gregory, D. (1984). The radical answer. Food Flav. Ingred. Proc. 6:18-25.

26. Olcott, H. S. (1967). Antioxidants. In Fish Oils, pp. 164-70. M. E. Stansby (ed.), Avi Publishing Co., Inc., Westport, Conn.

27. Einset, E., H. S. Olcott, and M. E. Stansby (1957). Oxidative deterioration in fish and fishery products. IV. Progress on studies concerning oxidation of extracted oils. Com. Fisheries Rev. 19(5a): 35-37.

28. Eisenbach, W. (1984). Supercritical fluid extraction: a film demonstration. Ber. Bunsenges Phys. Chem. 88: 882-87.

29. Nelson, R. W. (1982). Liquid CO_2 extraction and fisheries research. Mar. Fish. Rev. 46(1): 28.

30. Crawford, R. V. (1983). The U. K. edible oils industry - change and challenge. Food Flav. Ingred. Proc. 6: 33-37.

31. Association of Official Analytical Chemists (1975). Official Methods of Analysis, 12th edition. AOAC, Washington.

32. Food and Agricultural Organization of the United Nations, Fishery Industries Division (1975). The production of fish meal and oil. FAO Fisheries Tech. Paper No. 142. Rome. 54 pp.

33. Stansby M. E. (1973). Problems discouraging use of fish oil in American manufactured shortening and margarine. J. Am. Oil Chem. Soc. 50:220A-25A.

34. Veranth, M. F. and K. Robe (1979). CO_2-enriched atmosphere keeps fish fresh more than twice as long. Food Processing, April, pp. 76-79.

35. Bergh, S. and R. McGinity (1979). Needed: full-service fishing ports. Food Engineering, March, pp. 96-100.

36. Haagsma, N. C.M. Van Gent, J.B. Luten, R.W. De Jong, and E. Van Doorn (1982). Preparation of w-3 fatty acid concentrate from cod liver oil. J. Am. Oil Chem. Soc. 59(3): 117-18.

37. Adlof, R.O. and E.A. Emken (1985). The isolation of omega-3 polyunsaturated fatty acids and methyl esters of fish oils by sliver resin chromatography. J. Am. Oil Chem. Soc. 62(11): 1192-95.

38. Rizvi, S.S.H., A.L. Benado, J.A. Zollweg, and J.A. Daniels (1986). Supercritical fluid extraction: fundamental principles and modeling methods. Food. Tech. 40(6):55-65.

39. Gruger, E.H.Jr. (1967). Fatty acid composition. In Fish Oil, pp.3-30. M.E. Stansby (ed.), Avi Publishing Co., Inc., Westport, Conn.

40. Stansby, M.E. (1979). Marine derived fatty acids or fish oils as raw material for fatty acids manufacture. J. Am. Oil Chem. Soc. 56:793A-96A.

10
Summary of the Health Implications of Dietary Fish and Fish Oils and Research Needs

HEALTH EFFECTS OF FISH OILS

In 1982, 755,600 people died from ischemic heart disease, 433,800 from cancer, and 157,700 from stroke [1]. Dietary fats appear to be involved in the etiology and progress of these diseases, and it is known that saturated fats and cholesterol precipitate and accentuate heart disease.

Dietary unsaturated fatty acids, however, reduce hyperlipidemia, a potent risk factor influencing heart disease. The strong advocacy for increased consumption of vegetable oils to reduce serum cholesterol, triglycerides, and low density lipoproteins has resulted in the consumption of high amounts of linoleic acid and other n-6 PUFAs. Recent evidence suggests that consumption of an excessive amount of n-6 PUFAs may predispose individuals to various pathophysiological conditions [2]. These include thrombosis, psoriasis, asthma, and arthritis in addition to enhanced tumor growth, metastasis, and cancer.

It is significant that the n-3 PUFAs of fish oils appear to be much more effective in reducing hyperlipidemia than vegetable oils; apparently, they are more effective inhibitors of fatty acid synthesis and lipoprotein formation in the liver and they enhance the catabolism of lipoproteins.

In addition, the n-3 PUFAs of fish may directly affect cardiovascular tone via their impact on platelet function [3].

Strokes caused by thrombosis and blockage of cerebral arteries are precipitated by the excessive production of thromboxane, which causes excessive platelet aggregation. Dietary n-3 PUFAs of fish oil significantly reduce platelet thromboxane synthesis without impairing prostacyclin synthesis. In fact, some of the n-3 PUFAs such as EPA are converted to a potent antiaggregatory prostacyclin isomer [10] (Table 10.1).

Cumulative evidence indicates that several cancers, particularly mammary tumors, are facilitated if not stimulated by dietary n-6 PUFAs, which are converted to PGE_2, a tumor growth-promoting prostaglandin. Dietary n-3 PUFAs effectively reduce the production of PGE_2 in animals and concomitantly reduce the growth of certain tumors [4].

There were approximately 3.2 thousand deaths from asthma in 1982 [1]. Recent evidence indicates that compounds generated by lipoxygenase in pulmonary tissue, such as leukotrienes and hydroxy fatty acids, are potent factors involved in asthma. By reducing the synthesis of these agents, dietary n-3 PUFAs may reduce the incidence of asthma [11].

TABLE 10.1 Fatalities in U.S. Population from Selected Causes in 1982

Major cardiovascular diseases	967,900
Diseases of the heart	755,600
Ischemic heart disease	555,200
Hypertensive heart disease	24,100
Cerebrovascular diseases	157,700
Asthma	3,100
Malignancies	422,100
Atherosclerosis	26,800

Source: Ref. 1

Summary of the Health Implications

TABLE 10.2 Selected Chronic Conditions in U.S. Population in 1982

Heart disease	16,926,000
Asthma	7,899
Dermatitis, eczema	8,652
Arthritis	30,207,000
Diabetes	5,767
Chronic Bronchitis	7,709

Source: Ref. 1

TABLE 10.3 Some Possible Mechanisms Whereby n-3 PUFAs May Affect Biochemical Phenomena Related to Eicosanoids.

* Displace arachidonic acid from phospholipids
* As free fatty acids, compete with linoleic acid for $\Delta 6$,-desaturase
* Compete with arachidonic acid for cyclooxygenase
* Produce acyl hydroperoxides, which are not as effective in activating cyclooxygenase
* Act as a source of PGI3 with antithrombotic properties
* Enhance synthesis of TXA3, a weak agonist which may reduce TXA2 binding
* Alter physical properties of membranes by increasing fluidity
* Alter phospholipase and/or acyltransferase activities
* Depress fatty acid synthesis
* Depress lipoprotein synthesis
* Enhance lipoprotein turnover

Psoriasis, which is manifested by neutrophil accumulation, enhanced blood flow, and abnormal arachidonic acid metabolism may be mediated by leukotrienes. Excessive localized production of leukotrienes attracts leukocytes via chemotaxis, which accumulate to cause psoriatic lesions [5]. It is conceivable that n-3 PUFAs in fish oil, by reducing the production of leukotrienes, can reduce the development of psoriasis.

The pain and discomfort associated with arthritis and inflammation may result from a localized overproduction of eicosanoids [2]. These same symptoms may reflect excessive dietary intake of n-6 PUFAs, which results in saturated tissue levels of arachidonic acid. Upon stimulation and/or perturbation, an excessive amount of arachidonic acid may be released, producing an excess of eicosanoids with hyperalgesic and chemotactic properties and resulting in discomfort, pain, and inflammation. Fish and/or fish oil n-3 PUFAs, by competing with arachidonic acid in tissue phospholipids, may effectively reduce the availability of the arachidonic acid precursor in the tissues and thereby reduce the production of eicosanoids.

Recent studies have shown that dietary fat composition and content of different fatty acids may contribute to the immune response by affecting membrane function and/or by enhancing immune cell synthesis of prostaglandins and leukotrienes [6]. Prostaglandins are involved in macrophage function and motility and act as intracellular messengers in a number of immunological responses. Leukotrienes activate T-lymphocytes. By affecting n-6 PUFA metabolism and immune cell synthesis of prostaglandins and leukotrienes, dietary n-3 PUFAs also may affect the immune response.

Unsaturated fats are also involved in the inflammatory response via the synthesis of leukotrienes. Leukotrienes exert chemokinetic and chemotactic properties and recruit leukocytes, mast cells, and macrophages to sites of injury or infection in the body. They also may be involved in atherogenesis via macrophage endothelial cell interactions [2].

For example, a localized perturbation or injury in the intima of an arterial wall may result in the production of eicosanoids including leukotrienes. These cause platelet adherence and aggregation. In addition, the leukotrienes and hydroxy fatty acids may recruit macrophages that adhere to and infiltrate into the tissue. The macrophages may ingest lipid protein particles and cholesterol and slowly become infested by an overgrowth of tissue stimulated by platelet-derived growth factors. n-3 PUFAs may play a therapeutic role in this scenario as well [7].

RESEARCH NEEDS

The dramatic health effects and implications of dietary fat and n-3 PUFAs stress the need for systematic dietary studies interrelating the effects of dietary fat content and composition and n-3 PUFA metabolism.

Because of the significant species differences in essential fatty acid metabolism and eicosanoid synthesis and metabolism, extrapolation of data from animal studies must be done with caution. The markedly different responses of platelets from various animal species to agonists may indicate that the impact of dietary n-3 PUFAs differs among species [8]. This possibility must be ascertained also before extrapolating data from animal to human observations.

Since different tissues and cells produce different amounts and types of eicosanoids, which in turn exert different physiological effects as activators or inhibitors of specific functions [9], detailed systematic studies are needed to determine the effect of dietary n-3 PUFAs on the composition and function of each eicosanoid produced. Little is known about the normal range of tissue eicosanoid concentrations and how they are affected by dietary fatty acids or wither n-3 or n-6 dietary fatty acid levels. questions concerning optimum intake, both amount and ratios, need to be answered [12].

Ultimately, comprehensive studies that include all facets of enzyme regulation must be conducted with humans. Essential fatty acid metabolism and eicosanoid synthesis should be carefully monitored in all organs and tissues. In addition, the effects of these and other dietary components, age, genetic background, and lifestyle need to be monitored in long-term studies.

The apparent beneficial effects of fish oils and n-3 PUFAs have accentuated the need for more basic information concerning the metabolic biochemistry of unsaturated fatty acids. Research to elucidate the kinetics of n-3 PUFA acylation/deacylation in specific metabolic pools of phospholipids and the manner in which these are affected by other dietary lipids, calcium, and various agonists is needed. The relationship between n-3 and n-6 PUFAs in various phospholipid pools and eicosanoid production in different tissues should be elucidated.

Research is needed to determine the interactions between n-3 and n-6 PUFAs and to establish optimum dosages of n-3 PUFAs in relation to total fat and n-6 PUFA intake. Studies should be designed to determine the relationship between the dose and frequency of n-3 PUFA consumption and n-3 PUFA turnover in different tissues.

Given the beneficial effects of n-3 PUFAs, an enormous amount of research is needed to carefully examine the possibility of supplementing nonseafood products with n-3 PUFAs to enhance dietary intake in the United States. The major challenge of n-3 PUFA supplementation will be to develop procedures and/or antioxidant agents that will stabilize these fatty acids against autoxidation in foods. The feasibility of microencapsulation should be carefully studied as a means of adding n-3 PUFAs to a wide range of food products.

Optimum dosage through fish consumption must also be coordinated with realistic dietary goals. Consideration should be given to fish oil supply and to alternative methods of obtaining n-3 PUFAs, such as through dietary supplementation or dietary enrichment with an n-3 PUFA concentrate.

The side effects of long-term fish and fish oil consumption also should be investigated, especially the possibility of altering vitamin E requirement with an increase in n-3 PUFA consumption. The presence of sterols and vitamins A and D in fish oils, particularly liver oils, may restrict the widespread use of fish oils as dietary supplements. Research is needed to accurately quantify the sterol content of marine oils and seafood and to determine the types and nature of sterols in the various oils.

New species of fish should be investigated as a dietary source of n-3 PUFAs, particularly underutilized species. Because fish oils vary greatly among species in EPA and DHA content, some species will be more effective sources of n-3 PUFAs than others.

Methods of fish oil production must be improved to guarantee the removal of pesticides and other contaminants, to prevent autoxidation and the production of undesirable flavors and odors, and to improve storage and packaging.

The growing success of aquacultural methods may increase the supply of and consumption of fish. Aquaculture may provide a controllable method to make fresh fish of good quality continually available in the marketplace. Increasing the n-3 PUFA content of fish raised by aquaculture should also be investigated. This would improve the nutritional and health value of seafood and may subsequently increase the demand for fish products.

Finally, nutritional education programs are needed to publicize the beneficial health effects of dietary fish and fish oil.

REFERENCES

1. U.S. Bureau of the Census (1985). Statistical Abstract of the United States. 1986. 106th ed. Washington, DC.

2. Lands. W.E.M. (1985). Fish and Human Health. Academic Press, New York.

3. Goodnight, S.H., W.S. Harris, W.E. Connor, and D.R. Illingworth (1982). Polyunsaturated fatty acids, hyperlipidemia, and thrombosis. Arteriosclerosis 2(2):87-113.

4. Karmali, R.A., J. Marsh, and C. Fuchs (1984). Effect of n-3 fatty acids on growth of a rat mammary tumor. J. Natl.Cancer Inst. 73:457-461.

5. Lee, T.H., R.L. Hoover, J.D. Williams, R.I. Sperling, J. Ravalese, B.W. Spur, D.R. Robinson, W. Corey, R.A. Lewis, K.F. Austen (1985). Effect of dietary enrichment with eicosapentaenoic acid docosahexaenoic acids on in vitro neutrophil and monocyte leukotriene generation and neutrophil function. New Eng.J.Med. 312:1217-24.

6. Johnson, P.V. and L.A. Marshall (1984). Dietary fat, prostaglandins, and the immune response. Progr.Food Nutr.Sci. 83:25.

7. Ross, R. (1984). Pathogenesis of plaque formation. Arteriosclerosis 4:323.

8. Morita, L., R. Takahashi, Y. Saito, and S. Muroto (1983). Effects of EPA on arachidonic acid metabolism in cultured vascular cells in platelets; species differences. Thromboxane Res. 31:211-17.

9. Kinsella, J.E., B. Lokesh, G. Bruckner, and B. German (1986) Effects of n-3 fatty acids on eicosanoids: possible tissue and species differences. J.Am.Oil Chem.Soc. 52(abstract).

10. von Schacky, C., S. Fischer, and P.C. Weber (1985). Lipids, platelet function, and eicosanoid formation in humans. J. Clin. Invest. 76:1626-31.

11. Lee, T.H., R.A. Lewis, D. Robinson, J.M. Drazen, and K.F. Austen (1984). The effects of a diet enriched in menhaden fish oil on the pulmonary response to antigen challenge. J. Allergy Clin. Immunol. 73:150.

12. Lands, W.E.M. (1986). Renewed questions about polyunsaturated fatty acids. Nutr. Rev. 44:189-95.

11
Potential Sources of Fish Oil: Fatty Fish in U.S. Waters

FATTY FISH IN U.S. WATERS

Many marine biologists contend that the number of fish species landed commercially is very restricted. As an example, only a small selection of fish species serve as sources of oil, and two species, the Atlantic and Gulf menhadens, account for the bulk of fish oil production in the United States. Many species of commercial importance have decreased in abundance as a result of overfishing [7]. More prudent use of the nations' marine and freshwater resources could be made by commercially exploiting underused fish species.

This chapter presents a listing of all fatty fish that inhabit U.S. waters. The listing includes species that are currently the principal sources of oil. In addition, it reviews those fish that have high contents of fat and/or n-3 PUFAs but that currently have little commercial or recreational value. Also included are fish whose commercial importance is based on their flesh but whose offals have a high fat content and could be a practical source of oil. This listing also may alert us to fatty fish whose numbers are greatly decreasing, some to near extinction, by man's actions.

The lipid content of fish is highly variable. Environmental factors such as diet, season, and temperature and biological differences such as age, sex, and size are known to affect lipid content and composition and to account for differences between and within species [8]. Species that do not migrate over wide areas and do not need to store fat as an energy source have relatively low fat contents [8]. Among members of a species, age and size, which determine the content of the diet, affect the fat level in the body. Furthermore, during the summer when food is plentiful fat contents may be elevated and then decline during the fall and winter months [9].

Tissues and organs in the body also differ in their fat content. Less than 1% of the lipids occur as phospholipids and are associated with the cellular structure of all tissues. Depending on the species, the majority of fat, which is in the form of triglycerides, may be dispersed throughout the flesh or be restricted to fat depots. In some fish, fat is stored in the liver; in others, it is concentrated immediately under the skin, beneath the lateral line, or along the top of the back. Fatty tissue also is located along the abdominal wall. The amount of fat dispersed in the flesh decreases from the head to the tail where the muscles used for swimming are located [8]. Any classification of fish into fatty and lean species must therefore be arbitrary.

Fish species are included in this listing if the mean fat content of their raw muscle, offal, or roe is equal to or greater than 4.5%, as indicated by the most complete data bank on the chemical and nutritional composition of seafood available [10]. The Atlantic and Gulf menhadens are the only species included for which information on the mean fat content of raw muscle, offal, or roe was not available. Atlantic and Pacific cod are included, even though the mean fat content of their muscle, offal (total), and roe is less than 4.5%, because their livers historically have been an important source of oil.

The fish species are arranged in alphabetical order by their common names. For each species, information on EPA and DHA contents, habitat, commercial and recreational importance, and current uses is given in the form of a concise outline.

Information on the n-3 PUFA content of some of the species was not found after an exhaustive literature search. In other cases, the data were taken from analyses of flesh or fillets, while data on

TABLE 11.1 The Proximate and Fatty Acid Composition of Fish Tissue

Average Nutrient and Fatty Acid Content (grams/100g)	Finfish Species			
	European Anchovy	Largemouth Bass	Rock Bass	Striped Bass
Protein	21.197		18.280	17.730
Fat (total lipid)	4.841	1.300	0.780	2.328
Carbohydrate				
Ash			1.040	1.040
Cholesterol (mg)			50.000	80.000
Saturated fatty acids	1.420	0.332	0.214	0.582
Myristoleic (14:1)	0.048			0.004
Palmitoleic (16:1)	0.443	0.125	0.073	0.173
Oleic (18:1n-9)	0.691	0.237	0.145	0.514
Gadoleic (20:1n-9)		0.027		0.066
Erucic (22:1n-9)	0.127			
Linoleic (18:2n-6)	0.108	0.040	0.016	0.017
Linolenic (18:3n-3)		0.042	0.017	0.017
Moroctic (18:4n-3)	0.061	0.017		
Arachidonic (20:4n-6)	0.008	0.082	0.068	
Timnodonic (20:5n-3)	0.596	0.067	0.035	0.194
Clupanodonic (22:5n-3)	0.032	0.081	0.044	
Docosahexaenoic (22:6n-3)	1.009	0.225	0.168	0.671
Total monounsaturated fatty acids	1.309	0.389	0.218	0.757
Total polyunsaturated fatty acids	1.813	0.554	0.349	0.899

Table 11.1 (continued)

Average Nutrient and Fatty Acid Content (grams/100g)	Finfish Species			
	White Bass	Burbot	Capelin	Carp
Protein	18.612	19.147	13.908	17.843
Fat (total lipid)	4.001	0.780	8.787	
Carbohydrate	0.233			
Ash	1.723	1.199	1.992	1.455
Cholesterol (mg)	68.000	59.500		67.183
Saturated fatty acids	0.953	0.211	1.767	1.195
Myristoleic (14:1)			0.040	0.055
Palmitoleic (16:1)	0.468	0.033	1.093	0.722
Oleic (18:1n-9)	1.212	0.127	0.817	1.268
Gadoleic (20:1n-9)	0.070	0.007	1.445	0.078
Erucic (22:1n-9)		0.004	1.535	0.443
Linoleic (18:2n-6)	0.107	0.012	0.065	0.570
Linolenic (18:3n-3)	0.136		0.057	0.298
Moroctic (18:4n-3)	0.049		0.152	0.064
Arachidonic (20:4n-6)	0.177	0.124	0.039	0.168
Timnodonic (20:5n-3)	0.292	0.091	0.662	0.262
Clupanodonic (22:5n-3)	0.103	0.034	0.073	0.090
Docosahexaenoic (22:6n-3)	0.435	0.124	0.530	0.126
Total monounsaturated fatty acids	1.750	0.171	4.930	2.567
Total polyunsaturated fatty acids	1.298	0.384	1.577	1.577

Table 11.1 (continued)

Average Nutrient and Fatty Acid Content (grams/100g)	Finfish Species			
	Brown Bullhead Catfish	Channel Catfish	Atlantic Cod	Pacific Cod
Protein	18.600	18.167	17.808	17.898
Fat (total lipid)	2.659	4.264	0.748	0.629
Carbohydrate				
Ash	1.100	1.263	1.175	1.199
Cholesterol (mg)	75.000	58.000	42.983	37.000
Saturated fatty acids	0.659	1.093	0.194	0.116
Myristoleic (14:1)		0.017	0.002	
Palmitoleic (16:1)	0.384	0.184	0.021	0.044
Oleic (18:1n-9)	0.715	1.532	0.093	0.072
Gadoleic (20:1n-9)	0.031	0.059	0.017	
Erucic (22:1n-9)		0.002	0.005	
Linoleic (18:2n-6)	0.122	0.438	0.007	0.009
Linolenic (18:3n-3)	0.150	0.043	0.003	0.003
Moroctic (18:4n-3)		0.037	0.003	
Arachidonic (20:4n-6)	0.136	0.114	0.037	0.024
Timnodonic (20:5n-3)	0.200	0.144	0.113	0.114
Clupanodonic (22:5n-3)	0.067	0.069	0.014	0.006
Docosahexaenoic (22:6n-3)	0.195	0.270	0.211	0.191
Total monounsaturated fatty acids	1.129	1.794	0.138	0.116
Total polyunsaturated fatty acids	0.871	1.114	0.389	0.347

Table 11.1 (continued)

Average Nutrient and Fatty Acid Content (grams/100g)	Finfish Species			
	Atlantic Croaker	Spiny Dogfish	Dolphin-fish	Black Drum
Protein	17.783	16.550	18.500	16.281
Fat (total lipid)	3.167	10.186	0.700	2.540
Carbohydrate		2.271		1.584
Ash	1.114	1.466		0.910
Cholesterol (mg)	61.000	51.800		
Saturated fatty acids	1.226	2.370	0.232	0.842
Myristoleic (14:1)	0.006	0.024		0.008
Palmitoleic (16:1)	0.523	0.641	0.065	0.355
Oleic (18:1n-9)	0.668	2.507	0.101	0.453
Gadoleic (20:1n-9)	0.097	0.630	0.006	0.124
Erucic (22:1n-9)		0.735	0.015	
Linoleic (18:2n-6)	0.052	0.198	0.009	0.028
Linolenic (18:3n-3)	0.010	0.066	0.007	0.010
Moroctic (18:4n-3)	0.013	0.075	0.007	0.023
Arachidonic (20:4n-6)	0.013	0.256	0.002	0.160
Timnodonic (20:5n-3)	0.139	0.815	0.051	0.118
Clupanodonic (22:5n-3)	0.097	0.260	0.015	0.098
Docosahexaenoic (22:6n-3)	0.110	1.313	0.139	0.170
Total monounsaturated fatty acids	1.293	4.537	0.186	0.939
Total polyunsaturated fatty acids	0.523	2.982	0.231	0.607

Table 11.1 (continued)

Average Nutrient and Fatty Acid Content (grams/100g)	Finfish Species			
	Freshwater Drum	Flounder	Yellowtail Flounder	Red Grouper
Protein	17.535	15.575	21.056	18.919
Fat (total lipid)	4.941	1.037	1.200	0.780
Carbohydrate		0.570		
Ash	1.084	1.207	1.232	1.160
Cholesterol (mg)	64.000	46.180		
Saturated fatty acids	1.239	0.311	0.351	0.242
Myristoleic (14:1)			0.005	
Palmitoleic (16:1)	0.876	0.081	0.091	0.032
Oleic (18:1n-9)	1.327	0.199	0.133	0.108
Gadoleic (20:1n-9)	0.089	0.070	0.032	
Erucic (22:1n-9)	0.132	0.044	0.001	
Linoleic (18:2n-6)	0.172	0.006	0.012	0.022
Linolenic (18:3n-3)	0.127	0.023	0.011	
Moroctic (18:4n-3)			0.020	
Arachidonic (20:4n-6)	0.248	0.044	0.053	0.059
Timnodonic (20:5n-3)	0.254	0.114	0.131	0.026
Clupanodonic (22:5n-3)	0.159	0.036	0.059	0.018
Docosahexaenoic (22:6n-3)	0.318	0.110	0.138	0.203
Total monounsaturated fatty acids	2.425	0.394	0.262	0.141
Total polyunsaturated fatty acids	1.278	0.332	0.423	0.327

Table 11.1 (continued)

Average Nutrient and Fatty Acid Content (grams/100g)	Finfish Species			
	Jewfish Grouper	Haddock	Atlantic Hake	Pacific Hake
Protein		18.791	18.422	16.879
Fat (total lipid)	1.300	3.973	0.640	1.617
Carbohydrate			5.152	
Ash		1.209	1.435	1.137
Cholesterol (mg)	49.000	62.933		
Saturated fatty acids	0.396	0.970	0.237	0.326
Myristoleic (14:1)		0.010		0.003
Palmitoleic (16:1)	0.044	0.125	0.056	0.113
Oleic (18:1n-9)	0.288	0.498	0.179	0.198
Gadoleic (20:1n-9)	0.039	0.072		0.031
Erucic (22:1n-9)	0.010	0.173	0.022	
Linoleic (18:2n-6)	0.014	0.071	0.009	0.034
Linolenic (18:3n-3)	0.013	0.019	0.008	0.035
Moroctic (18:4n-3)		0.021		0.025
Arachidonic (20:4n-6)	0.034	0.173	0.004	0.054
Timnodonic (20:5n-3)	0.052	0.443	0.059	0.267
Clupanodonic (22:5n-3)	0.011	0.143		0.048
Docosahexaenoic(22:6n-3)	0.398	0.943	0.031	0.277
Total monounsaturated fatty acids	0.381	0.878	0.257	0.344
Total polyunsaturated fatty acids	0.523	1.812	0.110	0.740

Table 11.1 (continued)

Average Nutrient and Fatty Acid Content (grams/100g)	Finfish Species			
	Red Hake	Silver Hake	Atlantic Halibut	Pacific Halibut
Protein	28.502	16.994	20.189	20.814
Fat (total lipid)	0.917	2.603	1.007	2.302
Carbohydrate			0.366	
Ash	1.419	1.194	1.298	1.357
Cholesterol (mg)			41.000	32.000
Saturated fatty acids	0.220	0.606		0.376
Myristoleic (14:1)		0.006		0.004
Palmitoleic (16:1)	0.046	0.224		0.188
Oleic (18:1n-9)	0.300	0.437		0.417
Gadoleic (20:1n-9)		0.088		0.143
Erucic (22:1n-9)		0.012		0.118
Linoleic (18:2n-6)	0.011	0.056		0.036
Linolenic (18:3n-3)		0.101		0.075
Moroctic (18:4n-3)	0.007	0.063		0.046
Arachidonic (20:4n-6)	0.041	0.201		0.160
Timnodonic (20:5n-3)	0.113	0.239		0.083
Clupanodonic (22:5n-3)	0.036	0.044		0.109
Docosahexaenoic (22:6n-3)	0.130	0.325		0.338
Total monounsaturated fatty acids	0.346	0.767		0.871
Total polyunsaturated fatty acids	0.338	1.029		0.847

Table 11.1 (continued)

Average Nutrient and Fatty Acid Content (grams/100g)	Finfish Species			
	Greenland Halibut	Atlantic Herring	Pacific Herring	Atlantic Mackerel
Protein	14.367	17.797	16.394	18.599
Fat (total lipid)	13.841	9.042	13.878	13.891
Carbohydrate	0.518			2.611
Ash	1.000	1.486	2.367	1.347
Cholesterol (mg)	46.078	60.050	76.633	80.000
Saturated fatty acids	2.621	2.226	3.529	3.532
Myristoleic (14:1)	0.018	0.041	0.056	0.009
Palmitoleic (16:1)	1.910	0.679	1.145	1.310
Oleic (18:1n-9)	2.593	1.654	3.156	4.504
Gadoleic (20:1n-9)	2.424	0.786	1.472	2.299
Erucic (22:1n-9)	2.132	0.916	1.617	3.413
Linoleic (18:2n-6)	0.133	0.142	0.208	0.425
Linolenic (18:3n-3)	0.046	0.113	0.062	0.207
Moroctic (18:4n-3)	0.143	0.233	0.269	0.371
Arachidonic (20:4n-6)	0.066	0.066	0.104	0.200
Timnodonic (20:5n-3)	0.570	0.774	1.050	1.620
Clupanodonic (22:5n-3)	0.096	0.060	0.187	0.433
Docosahexaenoic (22:6n-3)	0.426	0.940	0.747	2.465
Total monounsaturated fatty acids	9.077	4.076	7.446	5.912
Total polyunsaturated fatty acids	1.481	2.328	2.626	3.641

Table 11.1 (continued)

Average Nutrient and Fatty Acid Content (grams/100g)	Finfish Species			
	Chub Mackerel	Horse Mackerel	Japanese Horse Mackerel	Atlantic Menhaden
Protein	20.790	19.617	19.150	
Fat (total lipid)	11.466	4.100	7.812	15.500
Carbohydrate			4.867	
Ash	1.370	2.350	1.158	
Cholesterol (mg)	51.895	41.000	47.803	
Saturated fatty acids	3.319	1.311	2.653	6.352
Myristoleic (14:1)		0.029	0.016	
Palmitoleic (16:1)	0.611	0.319	0.605	1.227
Oleic (18:1n-9)	0.778	0.590	0.189	0.140
Gadoleic (20:1n-9)	0.778	0.266	0.189	0.140
Erucic (22:1n-9)	1.057	0.319	0.182	0.264
Linoleic (18:2n-6)	0.222	0.053	0.103	0.171
Linolenic (18:3n-3)	0.034	0.041	0.093	0.295
Moroctic (18:4n-3)	0.293	0.041	0.067	0.295
Arachidonic (20:4n-6)	0.121	0.033	0.115	0.186
Timnodonic (20:5n-3)	0.761	0.385	0.511	1.584
Clupanodonic (22:5n-3)	0.145	0.106	0.159	0.248
Docosahexaenoic (22:6n-3)	1.537	0.319	1.718	1.988
Total monounsaturated fatty acids	4.444	1.524	2.514	3.712
Total polyunsaturated fatty acids	3.113	0.979	2.213	4.613

Table 11.1 (continued)

Average Nutrient and Fatty Acid Content (grams/100g)	Finfish Species			
	Striped Mullet	Ocean Perch	Yellow Perch	White Perch
Protein	19.241	18.624	19.308	19.800
Fat (total lipid)	3.722	1.631	0.885	2.500
Carbohydrate				
Ash	1.194	1.198	1.296	1.200
Cholesterol (mg)	49.460	42.010	90.000	80.000
Saturated fatty acids	1.114	0.288	0.240	0.651
Myristoleic (14:1)	0.166	0.004		
Palmitoleic (16:1)	0.563	0.102	0.070	0.371
Oleic (18:1n-9)	0.825	0.245	0.081	0.664
Gadoleic (20:1n-9)	0.022	0.951		0.029
Erucic (22:1n-9)	0.031	0.271		
Linoleic (18:2n-6)	0.087	0.033	0.014	0.095
Linolenic (18:3n-3)	0.025	0.068	0.015	0.092
Moroctic (18:4n-3)	0.082	0.028	0.012	0.050
Arachidonic (20:4n-6)	0.096	0.004	0.068	0.134
Timnodonic (20:5n-3)	0.216	0.095	0.102	0.279
Clupanodonic (22:5n-3)	0.099	0.027	0.037	0.040
Docosahexaenoic(22:6n-3)	0.108	0.250	0.235	0.095
Total monounsaturated fatty acids	1.075	0.740	0.151	1.064
Total polyunsaturated fatty acids	0.713	0.506	0.483	0.785

Table 11.1 (continued)

Average Nutrient and Fatty Acid Content (grams/100g)	Finfish Species			
	Northern Pike	Walleye Pike	European Plaice	Pollock
Protein	19.255	19.137	18.000	18.437
Fat (total lipid)	0.687	1.215	1.500	20.242
Carbohydrate				
Ash	1.204	1.199	1.320	1.371
Cholesterol (mg)	39.000	86.000	70.000	70.600
Saturated fatty acids	0.164	0.306	0.390	3.561
Myristoleic (14:1)	0.005			0.024
Palmitoleic (16:1)	0.078	0.120	0.202	0.316
Oleic (18:1n-9)	0.109	0.241	0.304	1.765
Gadoleic (20:1n-9)	0.017		0.084	0.588
Erucic (22:1n-9)	0.007		0.040	0.264
Linoleic (18:2n-6)	0.044	0.032	0.016	0.229
Linolenic (18:3n-3)	0.029	0.017	0.010	
Moroctic (18:4n-3)			0.022	0.131
Arachidonic (20:4n-6)	0.039	0.072	0.042	0.674
Timnodonic (20:5n-3)	0.045	0.105	0.170	1.870
Clupanodonic (22:5n-3)	0.019	0.046	0.042	0.592
Docosahexaenoic (22:6n-3)	0.102	0.277	0.127	9.257
Total monounsaturated fatty acids	0.216	0.361	0.630	2.957
Total polyunsaturated fatty acids	0.278	0.548	0.429	12.752

Table 11.1 (continued)

Average Nutrient and Fatty Acid Content (grams/100g)	Finfish Species			
	Florida Pompano	Canary Rockfish	Sablefish	Atlantic Salmon
Protein	18.478	18.795	13.413	19.533
Fat (total lipid)	9.466	1.784	15.299	5.357
Carbohydrate				3.341
Ash	1.100	1.186	1.051	1.652
Cholesterol (mg)	50.000	34.500	49.000	
Saturated fatty acids	3.822	0.451	3.463	0.914
Myristoleic (14:1)			0.038	
Palmitoleic (16:1)	0.614	0.119	1.323	0.234
Oleic (18:1n-9)	2.204	0.376	4.409	1.259
Gadoleic (20:1n-9)		0.025	1.537	0.208
Erucic (22:1n-9)		0.014	0.103	0.275
Linoleic (18:2n-6)	0.132	0.029	0.178	0.161
Linolenic (18:3n-3)		0.014	0.103	0.275
Moroctic (18:4n-3)		0.023	0.124	0.078
Arachidonic (20:4n-6)	0.268	0.036	0.113	0.249
Timnodonic (20:5n-3)	0.190	0.211	0.733	0.299
Clupanodonic (22:5n-3)	0.220	0.029	0.183	0.267
Docosahexaenoic (22:6n-3)	0.428	0.314	0.777	1.039
Total monounsaturated fatty acids	2.818	0.535	8.718	1.961
Total polyunsaturated fatty acids	1.238	0.657	2.211	2.367

Table 11.1 (continued)

Average Nutrient and Fatty Acid Content (grams/100g)	Finfish Species			
	Chinook Salmon	Chum Salmon	Coho Salmon	Pink Salmon
Protein	28.519	20.137	21.615	19.937
Fat (total lipid)	10.444	6.639	5.974	3.447
Carbohydrate				
Ash	1.367	1.180	1.206	1.222
Cholesterol (mg)		74.000		
Saturated fatty acids	2.729	1.679	1.216	0.626
Myristoleic (14:1)			0.026	
Palmitoleic (16:1)	0.962	0.516	0.593	0.174
Oleic (18:1n-9)	3.042	1.707	1.323	0.612
Gadoleic (20:1n-9)	0.491	0.362	0.207	0.139
Erucic (22:1n-9)	0.376	0.630	0.130	0.122
Linoleic (18:2n-6)	0.115	0.126	0.282	0.056
Linolenic (18:3n-3)	0.094	0.067	0.208	0.038
Moroctic (18:4n-3)	0.157	0.134	0.145	0.101
Arachidonic (20:4n-6)	0.167	0.060	0.156	0.087
Timnodonic (20:5n-3)	0.857	0.476	0.361	0.470
Clupanodonic (22:5n-3)	0.251	0.154	0.235	0.108
Docosahexaenoic (22:6n-3)	0.617	0.608	0.537	0.657
Total monounsaturated fatty acids	4.872	3.216	2.279	1.047
Total polyunsaturated fatty acids	2.258	1.626	1.923	1.517

Table 11.1 (continued)

Average Nutrient and Fatty Acid Content (grams/100g)	Finfish Species			
	Sockeye Salmon	American Sandlance	Sand Seatrout	Spotted Seatrout
Protein	21.300		18.094	20.231
Fat (total lipid)	8.556	7.200	2.325	1.700
Carbohydrate			0.441	
Ash	1.177		0.990	1.245
Cholesterol (mg)				
Saturated fatty acids	1.633	1.757	0.824	0.531
Myristoleic (14:1)	0.043	0.010		
Palmitoleic (16:1)	0.571	0.920	0.395	0.184
Oleic (18:1n-9)	1.508	0.645	0.516	0.238
Gadoleic (20:1n-9)	1.457	0.700	0.028	
Erucic (22:1n-9)	0.925	0.801		
Linoleic (18:2n-6)	0.415	0.091	0.005	0.042
Linolenic (18:3n-3)	0.100	0.056	0.002	
Moroctic (18:4n-3)	0.111	0.192	0.005	
Arachidonic (20:4n-6)	0.103	0.038	0.065	0.112
Timnodonic (20:5n-3)	0.566	0.888	0.072	0.094
Clupanodonic (22:5n-3)	0.043	0.047	0.047	0.042
Docosahexaenoic(22:6n-3)	0.713	0.530	0.280	0.100
Total monounsaturated fatty acids	4.504	3.075	0.938	0.421
Total polyunsaturated fatty acids	2.051	1.841	0.476	0.390

Table 11.1 (continued)

Average Nutrient and Fatty Acid Content (grams/100g)	Finfish Species			
	Rainbow Smelt	Red Snapper	European Sole	Sprat
Protein	17.224	21.742	17.700	17.006
Fat (total lipid)	2.550	1.151	1.175	5.755
Carbohydrate				0.634
Ash	1.551	1.296		1.100
Cholesterol (mg)	70.000		50.492	38.500
Saturated fatty acids	0.542	0.296	0.421	1.523
Myristoleic (14:1)	0.008	0.005		0.021
Palmitoleic (16:1)	0.240	0.063	0.186	0.706
Oleic (18:1n-9)	0.489	0.172	0.248	1.040
Gadoleic (20:1n-9)	0.033	0.022		
Erucic (22:1n-9)	0.001		0.069	0.484
Linoleic (18:2n-6)	0.054	0.020	0.006	0.095
Linolenic (18:3n-3)	0.059	0.005	0.008	
Moroctic (18:4n-3)	0.031	0.010		0.049
Arachidonic (20:4n-6)	0.066	0.052	0.018	0.023
Timnodonic (20:5n-3)	0.330	0.054	0.008	0.543
Clupanodonic (22:5n-3)	0.022	0.068	0.068	0.059
Docosahexaenoic (22:6n-3)	0.502	0.270	0.096	0.910
Total monounsaturated fatty acids	0.771	0.262	0.503	2.250
Total polyunsaturated fatty acids	1.063	0.479	0.204	1.679

Table 11.1 (continued)

Average Nutrient and Fatty Acid Content (grams/100g)	Finfish Species			
	White Sucker	Pumpkinseed Sunfish	Swordfish	Rainbow Trout
Protein	16.827	19.400	19.645	20.545
Fat (total lipid)	2.322	0.700	3.855	3.358
Carbohydrate				3.319
Ash	1.585	1.100	1.380	1.303
Cholesterol (mg)	43.000	67.000	36.680	56.667
Saturated fatty acids	0.520	0.191	1.189	0.728
Myristoleic (14:1)	0.008			
Palmitoleic (16:1)	0.442	0.057	0.272	0.178
Oleic (18:1n-9)	0.348	0.096	1.178	0.694
Gadoleic (20:1n-9)	0.017	0.007	0.132	0.143
Erucic (22:1n-9)			0.093	0.147
Linoleic (18:2n-6)	0.077	0.021	0.031	0.280
Linolenic (18:3n-3)	0.060	0.014	0.202	0.134
Moroctic (18:4n-3)	0.038			0.054
Arachidonic (20:4n-6)	0.118	0.108	0.074	0.128
Timnodonic (20:5n-3)	0.219	0.051	0.117	0.155
Clupanodonic (22:5n-3)	0.076	0.046		0.113
Docosahexaenoic (22:6n-3)	1.219	0.099	0.575	0.484
Total monounsaturated fatty acids	0.815	0.160	1.675	1.162
Total polyunsaturated fatty acids	0.921	0.339	0.999	1.347

Fatty Fish in U.S. Waters

Table 11.1 (continued)

Average Nutrient and Fatty Acid Content (grams/100g)	Finfish Species			
	Arctic Char Trout	Brook Trout	Lake Trout	Unspecified tuna
Protein	22.650	21.500	17.971	24.711
Fat (total lipid)	7.700	2.732	9.700	2.458
Carbohydrate			0.549	
Ash		1.300	1.038	1.431
Cholesterol (mg)		68.000	48.000	
Saturated fatty acids	1.703	0.744	1.886	1.006
Myristoleic (14:1)				
Palmitoleic (16:1)	1.457	0.295	0.891	0.187
Oleic (18:1n-9)	1.935	0.622	2.890	0.436
Gadoleic (20:1n-9)	0.563		0.136	
Erucic (22:1n-9)	1.025			0.053
Linoleic (18:2n-6)	0.054	0.166	0.443	0.043
Linolenic (18:3n-3)	0.023	0.171	0.393	
Moroctic (18:4n-3)	0.039	0.151	0.532	0.012
Arachidonic (20:4n-6)	0.039	0.151	0.532	0.012
Timnodonic (20:5n-3)	0.062	0.203	0.500	0.111
Clupanodonic (22:5n-3)	0.177	0.046	0.477	0.032
Docosahexaenoic (22:6n-3)	0.540	0.254	1.212	0.410
Total monounsaturated fatty acids	4.979	0.917	3.916	0.625
Total polyunsaturated fatty acids	0.933	1.071	3.686	0.613

Table 11.1 (continued)

Average Nutrient and Fatty Acid Content (grams/100g)	Finfish Species			
	Albacore Tuna	Bluefin Tuna	Skipjack Tuna	Yellowfin tuna
Protein	25.819	23.374	22.614	23.375
Fat (total lipid)	4.905	6.592	1.852	0.837
Carbohydrate		1.415	4.180	3.981
Ash	1.483	1.171	1.300	1.340
Cholesterol (mg)	54.000	38.233	46.940	44.650
Saturated fatty acids	1.367	1.872	0.699	0.273
Myristoleic (14:1)				
Palmitoleic (16:1)	0.245	0.243	0.076	0.029
Oleic (18:1n-9)	0.878	1.377	0.279	0.126
Gadoleic (20:1n-9)	0.139	0.412	0.037	0.008
Erucic (22:1n-9)	0.077	0.353	0.013	0.017
Linoleic (18:2n-6)	0.097	0.080	0.034	0.009
Linolenic (18:3n-3)	0.170		0.014	
Moroctic (18:4n-3)	0.092	0.059	0.009	0.006
Arachidonic (20:4n-6)	0.110	0.065	0.055	0.032
Timnodonic (20:5n-3)	0.376	0.423	0.151	0.043
Clupanodonic (22:5n-3)	0.031	0.187	0.028	0.015
Docosahexaenoic(22:6n-3)	1.104	1.327	0.395	0.210
Total monounsaturated fatty acids	1.339	2.385	0.404	0.180
Total polyunsaturated fatty acids	1.981	2.141	0.672	0.328

Table 11.1 (continued)

Average Nutrient and Fatty Acid Content (grams/100g)	Finfish Species		
	Lake Whitefish	Atlantic Wolffish	Whiting
Protein	19.020	17.500	19.889
Fat (total lipid)	5.968	2.390	0.500
Carbohydrate	1.146	5.020	
Ash	1.127	1.160	1.500
Cholesterol (mg)	60.000		31.000
Saturated fatty acids	1.015	0.417	0.164
Myristoleic (14:1)			
Palmitoleic (16:1)	0.582	0.216	0.033
Oleic (18:1n-9)	1.509	0.592	0.067
Gadoleic (20:1n-9)	0.116	0.130	0.004
Erucic (22:1n-9)	0.028	0.021	0.004
Linoleic (18:2n-6)	0.205	0.010	0.009
Linolenic (18:3n-3)	0.205	0.010	0.004
Moroctic (18:4n-3)	0.055	0.070	
Arachidonic (20:4n-6)	0.250	0.102	0.048
Timnodonic (20:5n-3)	0.355	0.352	0.055
Clupanodonic (22:5n-3)	0.183	0.050	0.016
Docosahexaenoic (22:6n-3)	1.054	0.362	0.096
Total monounsaturated fatty acids	2.235	0.959	0.109
Total polyunsaturated fatty acids	2.407	0.970	0.228

Source: Refs. 66 and 67.

TABLE 11.2 The Proximate and Fatty Acid Composition of Fish Oils.

Average Nutrient and Fatty Acid Content (grams/100g)	Finfish Species			
	Cod liver	Herring	Menhaden	MaxEPA
Protein				
Fat (total lipid)	100.000	100.000	100.000	100.000
Carbohydrate				
Ash				
Cholesterol (mg)	570	766	521	600
Saturated fatty acids	17.6	19.2	33.6	25.4
Myristoleic (14:1)				
Palmitoleic (16:1)				
Oleic (18:1n-9)				
Gadoleic (20:1n-9)				
Erucic (22:1n-9)				
Linoleic (18:2n-6)				
Linolenic (18:3n-3)	0.7	0.6	1.1	0
Moroctic (18:4n-3)				
Arachidonic (20:4n-6)				
Timnodonic (20:5n-3)	9.0	7.1	12.7	17.8
Clupanodonic (22:5n-3)				
Docosahexaenoic (22:6n-3)	9.5	4.3	7.9	11.6
Total monounsaturated fatty acids	51.2	60.3	32.5	28.3
Total polyunsaturated fatty acids	25.8	16.1	29.5	41.1

Table 11.2 (continued)

Average Nutrient and Fatty Acid Content (grams/100g)	Finfish Species	
	Salmon	Shark
Protein		
Fat (total lipid)	100.000	
Carbohydrate		
Ash		
Cholesterol (mg)	485	
Saturated fatty acids	23.8	
Myristoleic (14:1)		
Palmitoleic (16:1)		2.5
Oleic (18:1n-9)		
Gadoleic (20:1n-9)		
Erucic (22:1n-9)		
Linoleic (18:2n-6)		
Linolenic (18:3n-3)	1.0	1.8
Moroctic (18:4n-3)		0.4
Arachidonic (20:4n-6)		2.8
Timnodonic (20:5n-3)	8.8	6.5
Clupanodonic (22:5n-3)		1.2
Docosahexaenoic (22:6n-3)	11.1	19.7
Total monounsaturated fatty acids	39.7	
Total polyunsaturated fatty acids	29.9	

Source: Refs. 66, 67, and 68.

viscera were not available. Such data are needed to promote the commercial use of fish offal as a source of oil. In addition, some of the chemical determinations of fat content were made decades ago. Data produced using the latest analytical methods are needed.

Tables 11.1 and 11.2 summarize the average nutrient and fatty acid composition of fish tissue and fish oils of various finfish species.

COMMON FISH SPECIES: FAT AND n-3 PUFA CONTENT, LANDINGS, AND USES

ALEWIFE (Alosa pseudoharengus)

MEAN FAT CONTENT OF RAW MUSCLE: 6.8% (2.9-15.2%) [10]

n-3 PUFA CONTENT:
(whole fish)
EPA   8.22 wt % [11]
DHA   5.96 wt % [11]

HABITAT IN U.S. WATERS:
The alewife is an anadromous species which can be found in US Atlantic waters between the Carolinas and Maine as well as in coastal rivers and streams [12]. Landlocked populations are also known [13].

U.S. LANDINGS:
Commercial fishing, which is carried out off the eastern Atlantic states, is centered in the Chesapeake Bay region [12]. In commercial fishery, the alewife and the blueback herring (Alosa aestivalis) are treated as one species [14,15]. Their combined 1977 commercial landing was 62,575,000 lbs [15]. The 1983 harvest totaled 31,244,000 lbs [16]. The species is also caught for recreational purposes [17].
There has been a decline in alewife stock abundance since 1969 [12]. The reasons for this have not been conclusively determined, although the disruption of spawning due to pollution and dam construction is thought to be involved [18].

USES:
A commercial fishery for marine alewife and blueback herring has existed since colonial times [12]. In the 1800s and early 1900s, it was mainly salted for us as food [18]. In recent years, the major uses of both species have been as bait [18], pet food [18], and for reduction to fish meal [15,18] and oil [15]. Three reduction plants were in existance in 1977 [15]. Alewife scales are also used in jewelry manufacturing as a source of pearl essence [19].

Fatty Fish in U.S. Waters 263

AMBERJACK, GREATER (Seriola dumerii)

MEAN FAT CONTENT OF RAW MUSCLE: 8.7% (0.8-21.2%) [10]

HABITAT IN U.S. WATERS:
The amberjack is found from the Caribbean Sea to the Atlantic Ocean off Massachusetts [20].

U.S. LANDINGS:
The total quantity landed commercially in 1977 was 249,000 lbs. [15]. Recreational landings are also made [21].

USES:
The amberjack is used as a foodfish [19].

ANCHOVY (Engraulis mordax)

MEAN FAT CONTENT (RAW): 10.7% (3.5-24.6) [10]

HABITAT IN U.S. WATERS:
Anchovies are found in the Pacific Ocean off California, Oregon and Washington [22].

U.S. LANDINGS:
Commercial landings are made from California to Washington [15]. In 1977, 223,270,000 lbs. were landed [15]. The 1983 figure was 22,305,000 lbs [15]. No sport fishery exists [23].

USES:
Between 1948 and the mid-1960s, the catch was canned for human and pet food [23]. In 1966, legislation was passed allowing the processing into fish meal and oil [23]. By 1977 this was its primary use. 6,171,000 lbs. of oil were produced that year [15]. However, by 1983 its most important use was as bait. Sixteen million lbs. of anchovy landed that year were used for this purpose, while only 6.3 million lbs. were reduced to oil [15]. Minor uses of anchovy include human consumption, particularly as canned anchovy paste [22].

BASS, BLACK SEA (Centropristis striata)

MEAN FAT CONTENT OF RAW OFFAL: 7.2% (5.3-9.5%) [10]

HABITAT IN U.S. WATERS:
This species is found in Atlantic waters from Florida to Massachusetts [21].

U.S. LANDINGS:
The recreational harvest is much greater than the commercial one [24]. 5,970,000 lbs. were landed commercially in 1977; 3,982,000 lbs. in 1983 [16]. The species may be overharvested in the northern part of its range [24].

USE:
It is highly valued as a foodfish [25].

BASS, STRIPED (Morone saxatilis)

MEAN FAT CONTENT OF RAW OFFAL: 8.3% (4.4-12.1) [10]

n-3 PUFA CONTENT:
   (fillet)   (egg yolk)   (egg oil globule)
EPA 0.17 g/100 g [26] 8.6 + 3.8 wt % [27] 6.4 + 1.6 wt % [27]
DHA 0.47 g/100 g [26] 19.9 + 2.2 wt % [15] 10.0 + 1.5 wt % [27]

HABITAT IN U.S. WATERS:
The striped bass is found in Atlantic waters from Florida to Maine [28], in Gulf waters from Florida to Louisiana [13] and in rivers along the Gulf coast [28]. As a result of a stocking program in the late 1800s, it also occurs along the Pacific coast from California to Washington [28].

U.S. LANDINGS:
The species supports a commercial and a much larger recreational fishery [28]. Since the 1930s its abundance has fluctuated [28]. In 1977, 5,521,000 lbs. were landed commercially [15]; in 1983, the catch totaled 1,679,000 lbs. [16]

BLOATER (Coregonus hoyi)

MEAN FAT CONTENT OF RAW MUSCLE: 14.0% [10]

HABITAT IN U.S. WATERS:
The bloater is found in the Great Lakes region with the exception of Lake Erie. It has probably been extirpated from Lake Ontario [29].

USES:
It has little value as a sport or commercial fish [30,31].

BONITO, ATLANTIC (Sarda sarda)

MEAN FAT CONTENT OF RAW MUSCLE: 6.8% (2.6-10.2%) [10]

Fatty Fish in U.S. Waters 265

HABITAT IN U.S. WATERS:
The Atlantic bonito is found in waters off the Atlantic coast states [19].

U.S. LANDINGS:
Commercial landings are made between North Carolina and Massachusetts [15] (see Pacific Bonito). It is also caught by sport fishermen [17].

USES:
It is considered a good foodfish but has a limited market in the U.S. [19].

BONITO, PACIFIC (Sarda chiliensis)

MEAN FAT CONTENT OF RAW MUSCLE: 5.5% (1.5-11.5%) [10]

HABITAT IN U.S. WATERS:
It is found from California to Alaska [32].

U.S. LANDINGS:
The 1977 total commercial landing of S. chiliensis and S. sarda was 23,040,000 lbs [15]. In 1983, 8,065,000 lbs. were landed [15]. Recreational landings are also made [32].

USE:
There is only a limited market for this fish. Small amounts are sold fresh while the majority of the catch is canned [32].

BULLHEAD, BLACK (Ictalurus melas)

MEAN FAT CONTENT OF RAW OFFAL: 5.6% (5.4-5.7%) [10]

HABITAT IN U.S. WATERS:
This species is found in the Great Lakes, the St. Lawrence river region, and from Montana to the Appalachians. It is characteristic of ponds, pools in streams and rivers and swampy habitats throughout its range [29]. It is also a commonly stocked farm pond species [13].

USE:
It is considered an excellent table fish [13,30].

BUTTERFISH (Peprilus triacanthus)

MEAN FAT CONTENT OF RAW MUSCLE: 11.2% (5.1-17.3%) [10]
MEAN FAT CONTENT OF RAW OFFAL: 8.7% (4.6-13.0%) [10]

HABITAT IN U.S. WATERS:
Butterfish are found in the Atlantic from Florida to Maine [33].They are most abundant between southern New England and Cape Hatteras [33].

U.S. LANDINGS:
Commercial landings are made between Florida and New England [15]. In 1977, 2,905,000 lbs. were landed [15]; in 1983 the quantity rose to 10,601,000 lbs. [16] The species is also caught by anglers [25].

USE:
It is considered an excellent foodfish [25].

CARP   (Cyprianus carpio)

MEAN FAT CONTENT OF RAW MUSCLE: 5.9% (1.3-14.8%) [10]
MEAN FAT CONTENT OF RAW OFFAL: 14.3% (9.8-18.0%) [10]

n-3 PUFA CONTENT:
(fillet)
EPA 0.17 g/100 g [26]
DHA 0.07 g/100 g [26]

HABITAT IN U.S. WATERS:
Carp is found in freshwater bodies throughout the U.S. [13].

U.S. LANDINGS:
Commercial landings totaled 24,824,000 lbs. in 1977 [15]. Recreational landings are also made [34].

USES:
It is considered a pest by many game fishermen and has been the object of numerous eradication programs[14,31]. Nevertheless, it is a popular gamefish in some parts of the U.S. [13] and is consumed as food [34]. One oil reduction plant was in existance in 1977 [15].

CATFISH, CHANNEL   (Ictalurus punctatus)

MEAN FAT CONTENT OF RAW MUSCLE: 5.2% (0.7-11.0%) [10]

n-3 PUFA CONTENT:
        (muscle)
EPA 0.79% FA (0.23-2.50% FA) [34]
DHA 1.87% FA (0.56-6.14% FA) [34]

HABITAT IN U.S. WATERS:
Catfish are found in freshwater bodies throughout the U.S. [29].They are characteristic of clear, medium to large rivers with

Fatty Fish in U.S. Waters

swift currents over sand or gravel-rock bottoms [29] and are also cultured in rearing ponds [31].

U.S. LANDINGS:
In 1977, 38,404,000 lbs. of catfish and bullheads were landed comercially [15]. Recreational landings are also made [31,36].

USES:
As a foodfish, catfish are consumed fresh or smoked [30,35].

CHUB, UTAH (Gila atraria)

MEAN FAT CONTENT OF RAW MUSCLE: 4.8% [10]
MEAN FAT CONTENT OF RAW OFFAL: 20.2% [10]

HABITAT IN U.S. WATERS:
The native range of the Utah chub is the waters of the ancient Lake Bonneville basin in Utah, Idaho, Wyoming and Nevada. It is also native to the Snake river drainage above Shoshone Falls. It has been introduced into a number of river systems outside its natural range [29].

USE:
It is used as bait [13].

CISCO, LAKE (Coregonus artedii)

MEAN FAT CONTENT OF RAW OFFAL: 7.2% [10]

n-3 PUFA CONTENT:
	(fillet)	(fillet)
EPA	5.9 wt % FA [37]	0.13 g/100 g [26]
DHA	13.3 wt % FA [37]	0.29 g/100 g [26]

HABITAT IN U.S. WATERS:
The lake cisco is widespread in the upper Mississippi river and Great Lake regions. It occurs mainly in lakes but can also be found in large rivers [29].

U.S. LANDING:
Commercial landings are made [30]. The 1977 harvest totaled 323,000 lbs.[15]. Pollution and predation by the sea lamprey have caused a continuing decline of the harvest [17].

USE:
The lake cisco is considered an excellent foodfish and is marketed filleted or smoked [30].

CISCO, LONGJAW   (Coregonus alpenae)

MEAN FAT CONTENT OF RAW MUSCLE: 10.3% (10.0-10.8%) [10]
MEAN FAT CONTENT OF RAW OFFAL: 21.5% [10]

HABITAT IN U.S. WATERS:
It was common in the deep waters of Lakes Michigan, Huron and Erie but is now thought to be extinct [29].

CISCO, SHORTJAW   (Coregonus zenithicus)

MEAN FAT CONTENT OF RAW MUSCLE: 9.0% [10]

HABITAT IN U.S. WATERS:
It was previously common in the Great Lakes, with the possible exception of Lake Ontario. Now populations appear to only exist in Lake Superior [29]. Its disappearance was probably due to overfishing [31].

USES:
It was important as a foodfish [30].

CISCO, SHORTNOSE   (Coregonus reighardi)

MEAN FAT CONTENT OF RAW MUSCLE: 13.1% [10]

HABITAT IN U.S. WATERS:
In the past, it was common in Lakes Michigan, Huron, and Ontario. The populations in Lake Michigan and Ontario have apparently been extirpated [29].

USES:
It is an excellent foodfish but the species is too rare to have much commercial value [30].

COBIA   (Rochycentron canadus)

MEAN FAT CONTENT OF RAW MUSCLE: 5.4% (3.8-10.0%) [10]

HABITAT IN U.S. WATERS:
The cobia is common in Atlantic waters south of the Chesapeake Bay. Occasionally, it is found as far north as Cape Cod [19]. The species also inhabits the Gulf of Mexico [20, 38].

U.S. LANDINGS:
The 1977 commercial catch totaled 115,000 lbs. [15]. Recreational landings make up a major part of the total catch [38].

USES:
It is marketed as food [25].

COD, ATLANTIC (Gadus morhua)

n-3 PUFA CONTENT:
       (liver oil)   (roe)
EPA 8.0 wt % [37] 9.5% [39]   9.9 wt % [40] 16.5 wt % [41]
DHA 14.3 wt % [37] 12.6% [39]   9.9 wt % [40] 22.1 wt % [41]

    (flesh)
EPA 17.6 wt % [41]
DHA 37.5 wt % [41]

HABITAT IN U.S. WATERS:
The species is found in the Atlantic Ocean from Maine to North Carolina. It is most numerous east and north of Cape Cod [42].

U.S. LANDINGS:
It is fished both comercially and for recreation throughout its range [42]. Commercial landings totaled 76,889,000 lbs. in 1977 [15]; 112,474,000 lbs. in 1983 [16]. In quantity, the 1970 recreational catch equaled 2/3 the commercial harvest [42].

USES:
It is primarily a foodfish [43], [17]

COD, PACIFIC (Gadus macrocephalus)

n-3 PUFA CONTENT:
    (liver)   (testes)
EPA 9.5 wt % [44] 12.2 wt % [44]
DHA 15.5 wt % [44] 18.7 wt % [44]

HABITAT:
The Pacific cod is found in waters between California and Alaska [43].

U.S. LANDINGS:
Commercial landings are made in Washington and Alaska [15]. In 1977, the catch totaled 10,706,000 lbs. [15]; in 1983, 108,990,000 lbs. [16].

USES:
It is used primarily as a foodfish [43] and is marketed fresh or frozen [45].

DRUM, FRESHWATER (Aplodinatus grunniens)

MEAN FAT CONTENT OF RAW MUSCLE: 5.5% (1.0-8.4%) [10]
MEAN FAT CONTENT OF RAW OFFAL: 7.2% [10]

n-3 PUFA CONTENT:
(fillet)
EPA   5.1 + 1.4 wt % [46]
DHA   6.9 + 4.2 wt % [46]

HABITAT IN U.S. WATERS:
This species has the greatest latitudinal range of any North American freshwater fish. It ranges over much of the U.S. east of the Rocky Mountains with the exception of the Atlantic slope [29].

U.S. LANDINGS:
Commercial, and to a lesser degree, recreational landings are made [17]. The 1977 commercial harvest totaled 5,729,000 lbs. [15].

USES:
The comercial catch is used mainly as feed on fur ranches [17]. and is sold as a foodfish in the Midwest [30].

EEL, COMMON (Anguilla rostrata)

MEAN FAT CONTENT OF RAW MUSCLE: 13.3% (11.6-14.9%) [10]

n-3 PUFA CONTENT:
(fillet)
EPA 0.34 g/100 g [26]
DHA 1.0 g/100 g [26]

HABITAT IN U.S. WATERS:
This is a catadromous species which spawns in the Sargasso Sea [17,31]. It is found in waters off the Atlantic coast states, in rivers of the eastern U.S., in the lower Mississippi river system and the Great Lakes [31].

U.S. LANDINGS:
Commercial landings are made from Florida to New England, in the Mississippi river and Lake Ontario [15]. During 1977, 2,102,000 lbs. were landed [15]. Sport fishery occurs in frozen estuaries [47].

USES:
The species has a diminishing commercial importance [25]. It is marketed fresh, smoked, pickled or jellied [19], [31], [17] and is also used as bait [47].

EEL, CONGER (Conger oceanicus)

MEAN FAT CONTENT OF RAW MUSCLE: 11.6% [10]

U.S. LANDINGS:
Commercial landings are made from Florida to New England [15]. 56,000 lbs. were caught during 1977 [15].

USE:
It is used as a foodfish [48].

EULACHON (Thaleichthys pacificus)

MEAN FAT CONTENT OF RAW MUSCLE: 6.5% [10]

HABITAT IN U.S. WATERS:
The smelt is an anadromous species found from California to Alaska [29,36].

U.S. LANDINGS:
It was once the source of a 600 ton/yr fishery; the commercial catch is now much smaller [49]. Recreational landings are also made [36].

USE:
This species was important to Indians as a source of oil and food. Dried and supplied with a wick, it was used for lighting [29]. Today it is used for bait and as feed on fur ranches [49]. It is also sold fresh for human consumption [36].

FLOUNDER, SUMMER (Paralichthys dentatus)

MEAN FAT CONTENT OF RAW OFFAL: 4.6% (2.2-6.9%) [10]

HABITAT IN U.S. WATERS:
It is most common in Atlantic waters between Florida and Cape Cod [47].

U.S. LANDINGS:
An important sport fishery exists [21]. The commercial catch is considerably smaller [47]. In 1975 it was estimated that the combined landings were greater than the estimated maximum sustainable yield [47].

HAKE, SQUIRREL (Urophycis chuss)

MEAN FAT CONTENT OF RAW OFFAL: 5.3% [10]

HABITAT IN U.S. WATERS:
This species is found in Atlantic waters from North Carolina to Maine [47].

U.S. LANDINGS:
The 1977 harvest totaled 3,870,000 lbs. [15]; that of 1983, 4,767,000 lbs. [16]. The species is not commonly sought by sport fishermen [17].

USE:
Historically, the hake has not had much demand as a food fish because of its soft meat and poor keeping qualities. In the 1940s it began to be sold as poultry and mink feed. It is now also marketed in the form of fresh or frozen fillets or fish cakes [47].

HALIBUT, ATLANTIC   (Hippoglossus hippoglossus)

MEAN FAT CONTENT OF RAW MUSCLE (DARK MEAT): 6.2% (3.9-8.5%) [10]

n-3 PUFA CONTENT:
   (fillet)
EPA 0.10 g/100 g [26]
DHA 0.30 g/100 g [26]

HABITAT IN U.S. WATERS:
It is found in Atlantic waters north of Virginia [21].

U.S. LANDINGS:
Commercial landings are made in New England [15]. 126,000 lbs. were caught in 1977 [15]. The species is no longer abundant enough to be of great importance in commercial fisheries [19].

USES:
It is used as a food fish [19]. The liver produces a vitamin-- rich oil [21].

HALIBUT, PACIFIC   (Hippoglossus stenolepsis)

MEAN FAT CONTENT OF RAW MUSCLE (DARK MEAT): 10.6% (5.7-16.7%) [10]

n-3 PUFA CONTENT:
   (fillet)   (oil)
EPA 0.11 g/100 g [26]   10.0 % [8]
DHA 0.20 g/100 g [26]   7.9 % [8]

HABITAT IN U.S. WATERS:
It is found in the Pacific between California and Alaska [32].

Fatty Fish in U.S. Waters 273

U.S. LANDINGS:
Commercial landings are made in Washington and Alaska [15]. During 1977, 17,310,000 lbs. were landed [15]. The total halibut (Atlantic and Pacific) catch during 1983 was 45,230,000 lbs; 99% of this was caught in the Pacific coast [16]. A reduced stock abundance and the consequent high fishing effort required plague management of the Pacific halibut [16].
RecreationaL landings are also made [45].

USE:
It is a food fish marketed frozen and smoked [50].

HERRING, ATLANTIC   (Clupea harengus harengus)

MEAN FAT CONTENT OF RAW MUSCLE: 9.8% (2.4-20.2%) [10]

n-3 PUFA CONTENT:
    (fillet)   (flesh)   (roe)
EPA 0.33 g/100 g [26]   8.6% [51]   11.9 wt % [40]   13.7 wt % [52]
DHA 0.58 g/100 g [26]   7.6% [51]   15.7 wt % [40]   31.4 wt % [52]

    (oil)
EPA 6.6 wt % [39]
DHA 6.8 wt % [39]

HABITAT IN U.S. WATERS:
It is found in the Atlantic from Cape Hatteras to Maine [53].

U.S. LANDINGS:
Commercial landings are made from New Jersey to New England [15]. 111,569,000 lbs. were landed in 1977 [15] and 51,262,000 lbs. in 1983 [16]. Decreased availability and reduced export demand may be contributing to the reduction in landing [16].

USE:
The smaller sized fish are canned as "sardines"; the larger sized ones are sold fresh, pickled or smoked [19]. One reduction plant producing oil and two producing meal were in existance in 1977 [15]. Including the products of the Pacific Sea Herring, 2000 tons of meal and 440,000 lbs. of oil were made that year [15].

HERRING, PACIFIC   (Clupea harengus pallasi)

MEAN FAT CONTENT OF RAW MUSCLE:8.5% (2.6-12.5%) [10]

n-3 PUFA CONTENT:
    (fillet)   (viscera)   (oil)
EPA 8.6 wt % [37]   0.76g/100g [26]   10.6 wt % [44]   8.6% [8]
DHA 7.6 wt % [37]   0.57g/100g [26]   3.8 wt % [44]   7.6% [8]

HABITAT IN U.S. WATERS:
It is found in the Pacific from California to Alaska [49].

U.S. LANDINGS:
Commercial landings are made off the Pacific coast states [15]. During 1977, 49,573,000 lbs. were landed; 41,229,000 lbs. were caught in 1983 [16].

USE:
A total of four reduction plants producing oil and meal were in operation in 1977 [15]. (see Atlantic Herring). As a foodfish, it is usually marketed salted, cured, or fresh [45]. It is also used as bait [45].

HERRING, ROUND (Etrumeus teres)

MEAN FAT CONTENT OF RAW MUSCLE (DARK MEAT): 9.61 [10]

HABITAT IN U.S. WATERS:
The species occurs from the Gulf of Mexico to Maine [25]; but is scarce north of New Jersey [25]. It may also be found in Pacific waters off California [45].

KIYI (Coregonus kiyi)

MEAN FAT CONTENT OF RAW MUSCLE: 8.7% [10]

HABITAT IN U.S. WATERS:
The kiyi is common in Lake Superior, extremely rare in Lakes Michigan and Huron, and is believed to have disappeared from Lake Ontario [29]. Some consider the apecies to be near extinction [31].

USE:
It has little commercial value [30].

LAMPREY, RIVER (Lampetra ayresi)

MEAN FAT CONTENT OF RAW MUSCLE: 13.0% [10]

HABITAT IN U.S. WATERS:
The river lamprey is a fish parasite present in fresh and saltwater bodies from California to Southern Alaska [29]. It feeds on flesh of herring and salmon [45].

LINGCOD (Ophiodon elongatus)

MEAN FAT CONTENT OF RAW OFFAL: 5.7% [10]

HABITAT IN U.S. WATERS:
It is found in Pacific waters from California to Alaska [49].

U.S. LANDINGS:
The 1977 commercial harvest totaled 5,895,000 [15]; that of 1983 reached 9,369,000 [16]. It is also highly regarded as a gamefish [49].

USE:
As a foodfish, it is marketed fresh [49].

MACKEREL, PACIFIC OR CHUB   (Scomber japonicus)

MEAN FAT CONTENT OF RAW MUSCLE: 6.6% (0.3-15.9%) [10]
MEAN FAT CONTENT OF RAW ROE: 5.2% [10]

n-3 PUFA CONTENT:
 (eviscerated)   (flesh)   (viscera)
EPA 1.1 g/100 g [26]   10.2 wt % [44]   11.2 wt % [44]
DHA 1.3 g/100 g [26]   7.3 wt % [44]   7.4 wt % [44]

HABITAT IN U.S. WATERS:
It is found off both the Atlantic [17] and Pacific coasts [21].

U.S. LANDINGS:
Commercial landings are made in California [15]. During 1977, 7,315,000 lbs. were landed5. This increased to 72,918,000 lbs. during 1983 [16]. It is also caught by sport fishermen [23].

USES:
It is used as a foodfish [17].

MACKEREL, ATLANTIC   (Scomber scombus)

MEAN FAT CONTENT OF RAW MUSCLE: 9.6% (0.7-24.0%) [10]
MEAN FAT CONTENT OF RAW OFFAL: 17.7% (15.5-19.4) [10]

n-3 PUFA CONTENT:
 (raw muscle)   (fillet)
EPA 6.5 % [10]   0.65 g/100g [26]   7.1 wt% [37]
DHA 12.9 % [10]   1.1 g/100g [26]   10.8 wt% [37]

HABITAT IN U.S. WATERS:
It is found in Atlantic waters between North Carolina to Maine [54]

U.S. LANDINGS:
Commercial landings are made from the Chesapeake Bay to Maine [15]. 3,042,000 lbs. were caught in 1977 [15] and 6,418,000 lbs. in

1983 [16]. During the 1980s there has been a decrease in the landings due to factors other than abundance, since the spawning stock size has been increasing [15]. It is also caught by sport fishermen [17].

USES:
As a food fish, it is consumed smoked, pickled, dried, and fresh [17].

MACKEREL, KING (Scomberomorus cavalla)

MEAN FAT CONTENT OF RAW MUSCLE: 4.6% [10]

HABITAT IN U.S. WATERS:
It is found in the Gulf of Mexico, and occasionally in the U.S. south Atlantic [38].

U.S. LANDINGS:
9,406,000 lbs. were harvested commercially in 1977 [15]; 6,688,000 lbs. in 1983 [16]. It is the most popular offshore sport fish in the Gulf of Mexico [38].

MACKEREL, SPANISH (Scomber maculatus)

MEAN FAT CONTENT OF RAW MUSCLE: 5.9% (0.6-14.4%) [10]
MEAN FAT CONTENT OF RAW OFFAL: 12.4 [10]

HABITAT IN U.S. WATERS:
The species is found in the southern Atlantic and in the Gulf of Mexico [38].

U.S. LANDINGS:
During 1977, 13,687,000 lbs. were landed commercially [15].- 6,142,000 lbs. represented the total catch during 1983 [16]. It is also a popular sport fish [20], [38].

USE:
It is a foodfish of high market value [19].

MENHADEN, ATLANTIC (Brevoortia tyrannus)

MEAN FAT CONTENT OF CANNED MUSCLE: 10.2% [10]

n-3 PUFA CONTENT:
(entire body)
EPA 10.2 wt % FA [37]
DHA 12.8 wt % FA [37]

Fatty Fish in U.S. Waters   277

HABITAT IN U.S. WATERS:
It occurs in Atlantic waters between northern Florida and Maine [21].

U.S. LANDINGS:
Commercial landings are made off the Atlantic coast states [15], [55]. The total catch (of both Atlantic and Gulf coast menhaden) was 1,796,182,000 lbs. during 1977, the greatest landing by volume and 34% of the total U.S. landing [15]. In 1983 a record 2,962,811,000 lbs. were caught [16]. The 1983 Atlantic coast landing was the largest since 1962. Spawning stock sizes have improved since the population decline in the 1960s but the magnitude and distribution of current fishing will not allow short-term landings from reaching much higher levels [15].

USE:
It is used primarily for reduction to meal, oil and solubles. In 1977, twenty-two reduction plants were in existence and 116,149,000 lbs. of oil were produced [15]. Small quantities are also used for bait and canned pet food [15], [16]. During 1983, 385,779,000 lbs. of oil were produced, accounting for 97% of the total U.S. fish oil production [16]. The species is held in low esteem as a food fish due to its extreme oiliness [25].

MENHADEN, GULF   (Brevoortia patronus)

MEAN FAT CONTENT OF RAW WHOLE FISH: 11.8% (2.1-17.8%) [10]

HABITAT IN U.S. WATERS:
It is found in the Gulf of Mexico [15].

U.S. LANDINGS:
(see Atlantic menhaden) It is believed the Gulf fishing will not be able to sustain the present levels of harvest and landings will eventually be reduced [15].

USES:
(see Atlantic menhaden)

MULLET, STRIPED   (Mugil cephalus)

MEAN FAT CONTENT OF RAW MUSCLE: 6.0% (0.7-20.2%) [10]
MEAN FAT CONTENT OF RAW OFFAL: 15.2% (1.2-26.7%) [10]

n-3 PUFA CONTENT:
(eviscerated)	(entire body, caught in Dec.)	(entire body, caught in July)
EPA 0.5 g/100 g [26]	7.5 wt. % [37]	11.8 wt. % [37]
DHA 0.39 g/100g [26]	13.4 wt. % [37]	3.2 wt. % [37]

HABITAT IN U.S. WATERS:
The species can be found in the Middle Atlantic states [15], the Chesapeake bay area [15], rivers in Louisiana and Texas [20], the Gulf of Mexico [20], and California [23]. It inhabits fresh to hypersaline waters, and shallow estuaries to deep waters [38].

U.S. LANDINGS:
Commercial landings are made from Texas to New Jersey [15], and off the shores of California [23]. The total cauch of all mullets was 27,657,000 lbs. in 1977 [15] and 25,549,000 lbs. in 1983 [16]. A moderate sport fishing also exists [23,38].

USE:
It is regarded as an excellent foodfish in Florida but not in the western Gulf of Mexico where it is not generally eaten [20]. Market conditions have also sometimes been poor in California [23]. The fish has also been used as bait [38].

PERCH, BLUE, WALLEYE   (Stizostedion vitreum glaucus)
MEAN FAT CONTENT OF RAW OFFAL: 8.8% (7.5-10.7%) [10]

HABITAT IN U.S. WATERS:
Its native range is in large streams, rivers, and lakes from Georgia and Alabama to North Dakota [29]. The species is considered endangered [29] or extinct in all of its range [17].

PERCH, OCEAN   (Sebastes marinus)

MEAN FAT CONTENT OF RAW OFFAL: 14.7 (0.4-25.0) [10]

n-3 PUFA CONTENT:
(fillet)
EPA   0.2   g/100 g [26]
DHA   0.18 g/100 g [26]

HABITAT IN U.S. WATERS:
The species occurs in Atlantic waters [21].

U.S. LANDINGS:
The 1977 commercial harvest totaled 35,000,000 lbs [15]. 13,289,000 lbs. were landed in 1983 [15].

USE:
It is used as a food fish [21].

PERCH, PACIFIC OCEAN   (Sebastes alutus)

MEAN FAT CONTENT OF RAW OFFAL: 7.8% [10]

n-3 PUFA CONTENT:
(entire edible portion)
EPA   9.3 wt % [37]
DHA 12.0 wt % [37]

HABITAT IN U.S. WATERS:
The species is found in Pacific waters from California to Alaska [21].

U.S. LANDINGS:
It has only recently begun to be exploited commercially [21]. 5,916,000 lbs. were caught in 1977 [15]; 5,987,000 lbs. in 1983 [16].

USE:
It is used as a foodfish and marketed as fillets [45].

PERCH, YELLOW   (Perca flavescens)

MEAN FAT CONTENT OF RAW OFFAL: 7.6% (5.0-9.2%) [10]

n-3 PUFA CONTENT:
       (fillet)
EPA   11.5 + 0.9 wt % [46]
DHA   26.4 + 2.0 wt % [46]

HABITAT IN U.S. WATERS:
It is found in fresh and brackish waters from South Carolina to Maine, and west of the Appalachians from Pennsylvania to Missouri and from Kansas to Montana [29]. The natural range has been enlarged through widespread stocking so that it is now present in many western states [13,29]. It is a very adaptable species that lives in a variety of habitats [29].

U.S. LANDINGS:
4,88,000 lbs. were harvested commercially in 1977 [15]. It is also highly esteemed as a game fish [13].

USE:
It is highly regarded as a foodfish [13.17].

PIKE, NORTHERN   (Esox lucius)

MEAN FAT CONTENT OF RAW OFFAL: 5.3% [10]

n-3 PUFA CONTENT:
     (flesh)   (fillet)   (testis)   (liver)
EPA 8.8% [56]   6.1 wt % [46] 7.3% [57]   4.4% [57]
DHA 12.3% [56] 30.1 wt % [46] 7.8% [57]   12.6% [57]

HABITAT IN U.S. WATERS:
The northern pike is found from New York, through the Great Lakes region to Nebraska [13]. It has also been widely introduced in many southern and western states [13]. It inhabits shallow, weedy areas of lakes, ponds and rivers [17].

U.S. LANDINGS:
Both commercial and recreational landings are made, although it is much more important as a gamefish tham as a commercial species [13,17,30].

USE:
The northern pike is available commercially in fresh and frozen forms [30].

PILCHARD (Sardinops caerulea)

MEAN FAT CONTENT OF RAW MUSCLE: 12.1% (3.1-15.6%) [10]

n-3 PUFA CONTENT:
  (oil)
EPA   16.9 % [58]
DHA   12.9 % [58]

HABITAT IN U.S. WATERS:
It is found in the Pacific Ocean from California to Alaska [23]. The species has all but disappeared; a moratorium on fishing declared in 1967 may have come too late to be effective in rebuilding the species [23].

U.S. LANDINGS:
Commercial landings are made from California to Washington; 5000 lbs. were caught in 1977 [15]. There is little sport fishing [23].

USE:
The bulk of the catch is canned [23].

POLLOCK (Pollachius virens)

MEAN FAT CONTENT OF RAW OFFAL: 8.1% (6.6-9.6%) [10]

n-3 PUFA CONTENT:
       (fillet)   (roe)
EPA   0.07 g/100 g [26]   11.5 wt % [52]
DHA   0.33 g/100 g [26]   27.7 wt % [52]

HABITAT IN U.S. WATERS:
It is found in Atlantic waters from New Jersey to Maine, and is occasionally seen off the Chesapeake Bay [13].

Fatty Fish in U.S. Waters

U.S. LANDINGS:
The 1977 total commercial harvest was 28,591,000 lbs. [15]; in 1983, 33,866,000 lbs. were caught [16]. The pollock is a very plentiful fish [17]; it has been stated that the catch could safely be increased about 85% [47]. It is also caught by sport fishermen [17].

USE:
Much of the catch is processed as high quality fillets [47]. It has become a substitute for the less abundant haddock [17].

RATFISH (Hydrolagus colliei)

MEAN FAT CONTENT OF RAW MUSCLE: 9.8% (0.7-18.1%) [10]

HABITAT IN U.S. WATERS:
The species is found in the Pacific Ocean from California to Alaska [49].

USE:
The flesh is edible but not considered desirable [49]. The liver oil is used as a lubricant for guns [21].

SABLEFISH (Anoploma fibria)

MEAN FAT CONTENT OF RAW MUSCLE: 15.2% [10]
MEAN FAT CONTENT OF RAW OFFAL: 14.7% (0.4-25.0) [10]

n-3 PUFA CONTENT:
    (fillet)   (entire edible portion) (body oil)
EPA 0.58 g/100 g [26]   8.5 % [37]   4.6 % [58]
DHA 0.61 g/100 g [26] 12.1 % [37]   2.2 % [58]

HABITAT IN U.S. WATERS:
The sablefish inhabits Pacific waters from California to Alaska [23].

U.S. LANDINGS:
Commercial landings are made off the Pacific coast states and Alaska [15]. Landings totaled 18,533,000 lbs. in 1977 and 40,151,000 lbs. in 1983 [16]. Recreational landings are also made, mostly of the younger fish [23].

USE:
As a foodfish, it is sold fresh [49], frozen [49], or smoked [23]. At times, the liver oil is in great demand [49].

SALMON, ATLANTIC (Salmo salar)

MEAN FAT CONTENT OF RAW MUSCLE: 5.5% (0.2-14.5%) [10]

n-3 PUFA CONTENT:
(fillet)   (flesh)   (liver)
EPA 0.18 g/100 g [26] 4.5 wt % [46] 3.4 wt % [40] 5.5 wt % [40]
DHA 0.61 g/100 g [26] 17.0 wt % [46] 2.5 wt % [40] 14.1 wt % [40]

HABITAT IN U.S. WATERS:
It is an anadromous species found in the Atlantic from the Connecticut river to Maine [29] but is uncommon south of Cape Cod [14]. A number of landlocked populations are also known [29]. Impassable dams and pollution have reduced or eliminated it in numerous locations [19], [25]. It is also in danger due to commercial overfishing in the Atlantic [25].

U.S. LANDINGS:
Commercial landings totaled less than 500 lbs. in 1977 [15]. It has greater value as a sport rather than a commercial fish [17].

USE:
It is a highly prized game- and foodfish [13,25].

SALMON, CHINOOK OR KING (Oncorhynchus tshawytscha)

MEAN FAT CONTENT OF RAW MUSCLE: 11.6% (2.2-19.0%) [10]

n-3 PUFA CONTENT:
(fillet)   (steak, ahead of
                 dorsal fin)
EPA 1.0 g/100 g [26]   8.2 wt. % [37]
DHA 0.72g/100 g [26]   5.9 wt. % [37]

HABITAT IN U.S. WATERS:
This is the least abundant of the Pacific salmon [29,36]. It is found from northern California to Alaska, having been extirpated from southern California [29]. Transplants to the Great Lakes have been successful [29,36].

U.S. LANDINGS:
It is the most commercially valuable salmon [36,49]. Commercial landings totaled 35,310,000 lbs. in 1977 [15] and 24,424,000 lbs. in 1983 [16]. Recreational landings are also made [23,36]. In recent years, the estimated ocean sport catch in California, Oregon and Washington has been close to 20% of the total landings [13].

USE:
It is consumed almost exclusively in the fresh state [23].

SALMON, CHUM OR KETA (Oncorhynchus keta)

MEAN FAT CONTENT OF RAW MUSCLE: 4.6% (1.3 - 7.9%) [10]
MEAN FAT CONTENT OF RAW ROE: 12.4% (10.2 - 17.0) [10]

n-3 PUFA CONTENT:
(steak, ahead of dorsal fin) (fillet)
EPA   6.7 wt % FA [37]   0.24 g/100 g [26]
DHA 16.1 wt % FA [37]   0.31 g/100 g [26]

HABITAT IN U.S. WATERS;
The species ranges from northern California to Alaska, in both rivers and coastal Pacific waters [29]. Transplanting efforts outside its natural range have been unsuccessful [29].

U.S. LANDINGS:
The 1977 commercial catch totaled 65,563,000 lbs. [15]; that of 1983, 79,920,000 lbs. [16]. It is only rarely taken by sport fishermen [13].

USE:
It is highly regarded as a foodfish.

SALMON, PINK (Oncorhynchus gorbuscha)

MEAN FAT CONTENT OF RAW MUSCLE: 4.8% (1.8 -12.5) [10]

n-3 PUFA CONTENT:
  (steak, ahead of
    dorsal fin)   (eggs)
EPA   13.5 wt % FA [37]   20.6 wt % FA [37]
DHA   18.9 wt % FA [37]   16.0 wt % FA [37]

HABITAT IN U.S. WATERS:
The pink salmon is found in coastal Pacific waters and rivers from California to Alaska [29]. It has also been successfully introduced into Lake Superior [30].

U.S. LANDINGS:
The commercial effort accounts for most of the harvest of this species [13]. In 1977, 143,645,000 lbs. were landed [15]; in 1983, the commercial catch was 194,140,000 lbs. [16].

USE:
It is highly regarded as a foodfish.

SALMON, RED OR SOCKEYE (Oncorhynchus nerka)

MEAN FAT CONTENT OF RAW MUSCLE: 6.9% (1.6-13.2%) [10]

n-3 PUFA CONTENT:
                      (hindpart
       (fillet)   flesh)   (head)
EPA 1.3 g/100 g [26]   4.9 wt % [44] 6.7 wt % [44]
DHA 1.7 g/100 g [26]   6.4 wt % [44] 4.7 wt % [44]

HABITAT IN U.S. WATERS:
It is found in California, Idaho, Oregon, Washington and Alaska [29]. Few transplants elsewhere have been successful [29].

U.S. LANDINGS:
It is the most abundant of the salmons, as evidenced by volume of commercial landings [15]. During 1977, 101,128,000 lbs. were caught [15]; the 1983 catch totaled 310,146,000 lbs [16].

USE:
As a foodfish, it is highly prized for its high oil content, excellent flavor and the color of its flesh [13].

SALMON, SILVER OR COHO (Oncorhynchus kisutch)

MEAN FAT CONTENT OF RAW MUSCLE: 5.7% (1.3-9.9%) [10]
MEAN FAT CONTENT OF RAW ROE: 11.2% (10.9-11.4%) [10]

n-3 PUFA CONTENT:
         (dorsal fin)   (Pacific, fillet)   (Lake Mich., fillet)
EPA   12.0 wt % [37]   0.82 g/100 g [26]   0.17 g/100 g [26]
DHA   13.8 wt % [37]   0.94 g/100 g [26]   0.34 g/100 g [26]

HABITAT IN U.S. WATERS:
It is found in Pacific Ocean waters and freshwater lakes from California to Alaska [29]. Since 1967, it is also caught in the Great Lakes [13].

U.S. LANDINGS:
The 1977 commercial harvest totaled 31,440,000 lbs [15]; 30,663,000 lbs. were landed in 1983 [16].

USE:
It is a foodfish consumed canned, smoked or fresh [49].

SAURY, ATLANTIC (Scomberesox saurus)

MEAN FAT CONTENT OF RAW MUSCLE: 6.1% (1.4-12.7%) [10]

HABITAT IN U.S. WATERS:
It inhabits the open waters of the Gulf of Mexico [20].

U.S. LANDINGS:
The species has no significant recreational or commercial value [21].

USE:
There is no market for this species [21].

SAURY, PACIFIC (Cololabis saira)

MEAN FAT CONTENT OF RAW MUSCLE: 11.1% (1.2-21.2%) [10]

HABITAT IN U.S. WATERS:
The species is found in the Pacific Ocean from California to Alaska [49].

U.S. LANDINGS:
Commercial landings are sporadic and unpredictable [23]. There is no significant sport fishing [23].

USE:
There is little demand for fresh or canned sauries [23].

SCAD (Trachurus lathami)

MEAN FAT CONTENT OF RAW MUSCLE: 11.4% (9.8-13.0%) [10]

HABITAT IN U.S. WATERS:
The scad inhabits the inshore waters of the eastern and Gulf coasts [20].

SCUP (Stenotomus chrysops)

MEAN FAT CONTENT OF RAW OFFAL: 12.1% (8.6-15.6%) [10]

HABITAT IN U.S. WATERS:
It is found in Atlantic waters from Florida to Maine [21].

LANDINGS:
Both commercial and recreational landings are made, particularly in the more northern parts of its range [21]. Between the early 1960s and 1970s, poor spawning success, and fishery pressure brought about a dramatic reduction in population size. Since then, there has been an increase in landings but it has not returned to its former abundance [59].

USE:
It is considered a good foodfish [21].

SEA TROUT, GRAY (Cynoscion regalis)

MEAN FAT CONTENT OF RAW OFFAL: 10.8% [10]

HABITAT IN U.S. WATERS:
It is found in Atlantic waters south of New England and occasionally as far north as Cape Cod [17].

U.S. LANDINGS:
Between 1910 and the late 1960s there was a general decline in the abundance of the species; since 1970, the population has begun to increase again [60]. The 1977 commercial landings totaled 18,927,000 lbs. [15]; in 1983, 17,543,000 lbs. were harvested [16]. An important sport fishery also exists [60].

USE:
It is highly regarded as a foodfish [25].

SHAD (Alosa sapidissima)

MEAN FAT CONTENT OF RAW MUSCLE: 11.2% (3.0-17.2%) [10]
MEAN FAT CONTENT OF RAW OFFAL: 16.3% (11.5-19.9%) [10]

HABITAT IN U.S. WATERS:
This anadromous species is found in the Atlantic coastal region from Florida to Maine [18]. It has been introduced to Pacific coast rivers [36].

U.S. LANDINGS:
Commercial landings are made from Florida to New England [15] and in the Columbia river in California [36]. 4,246,000 lbs. were harvested in 1977 [15]. Off the Atlantic states the catch has declined drastically from levels taken decades ago [18,61]. It has been suggested that this is due to pollution, the siltation of spawning rivers, dam construction, and overharvesting but there is insufficient data to determine the true causes [18].
Recreational landings are also made. In recent years, along the Atlantic coast they have equaled the commercial catch [61].

Fatty Fish in U.S. Waters   287

USE:
   In the past, it was considered a highly valuable food fish but in recent years the demand has declined substantially [18]. Roe is processed and sold as caviar and the flesh is sold fresh or smoked [17].

SHAD, GIZZARD   (Dorosoma cepedianum)

MEAN FAT CONTENT OF RAW MUSCLE: 11.2% (0.9-19.8%) [10]
MEAN FAT CONTENT OF RAW OFFAL: 22.8% (22.4-23.1) [10]

HABITAT IN U.S. WATERS:
   It is found in coastal fresh or brackish water from Texas to Cape Cod but is not common north of Virginia. It is also seen in the Great Lakes region [19] and in the Mississippi river and its drainage [13].

U.S. LANDINGS:
The 1977 commercial catch totaled 1,370,000 lbs. [15].

USE:
   The gizzard shad is not popular as a foodfish [30]. It is processed into oil, fertilizer, and animal feed [31] and is also used as bait [13].

SHAD, HICKORY   (Alosa mediocris)

MEAN FAT CONTENT OF RAW MUSCLE: 4.8% [10]

HABITAT IN U.S. WATERS:
It is found in coastal and freshwater bodies from Florida to Maine [13].

USES:
   The species is a popular gamefish in southern U.S. rivers [13]. It has no commercial or recreational importance north of the Chesapeake Bay [25].

SHARK, GRAYFISH   (Squalus acanthias)

MEAN FAT CONTENT OF RAW MUSCLE: 9.5% (4.8-15.3%) [10]
MEAN FAT CONTENT OF RAW ROE: 24.3% (23.5-25.0%) [10]

n-3 PUFA CONTENT:
   (flesh ahead of   (fillet)   (liver oil)
   dorsal fin)
EPA 7.9 wt. % FA [37]   0.66 g/100g [26]   3.7 wt. % FA [37]
DHA 10.4 wt. % FA [37] 1.3 g/100g [26]   6.5 wt. % FA [37]

HABITAT IN U.S. WATERS:
It is found in waters between Florida and Maine and California and Alaska [49]

U.S. LANDINGS:
Commercial landings are made in the Atlantic and Pacific coast states [15,49]. The total catch in 1977 was 7,827,000 lbs. [15]; in 1983 it reached 14,453,000 lbs. [16]. The species may only be able to sustain low fishing pressure over a long period because of its low reproductive potential [62].

USE:
The flesh is used in reduction plants for meal (used mainly in animal feed [63]) and oil [47,49]. It is a popular student laboratory animal [63]. The flesh is considered quite palatable but only small quantities are found in U.S. markets [63]. In recent years, small quantities of fillets have started to be exported from the Atlantic coast states to Europe [62].

SHARK, GREENLAND (Somniosus microcephalus)

MEAN FAT CONTENT OF RAW MUSCLE: 10.0% [10]

HABITAT IN U.S. WATERS:
The species occurs in Atlantic and Gulf of Mexico waters [21].

SHARK, SALMON (Lamna ditropsis)

MEAN FAT CONTENT OF RAW MUSCLE: 12.2% (0.2-24.2%) [10]

n-3 PUFA CONTENT:
(dorsal flesh)
EPA 0.05 g/100 g [26]
DHA 0.47 g/100 g [26]

HABITAT IN U.S. WATERS:
It is found in Pacific waters from California to Alaska [21].

U.S. LANDINGS:
This species is not classified as a game fish although it is caught by sport fishermen in coastal Washington [63].

SHARK, SEVENGILL (Notorynchus maculatus)

MEAN FAT CONTENT OF RAW MUSCLE: 13.1% [10]

HABITAT IN U.S. WATERS:
It is found in ocean waters from California to Washington [49].

USE:
This species has little current comercial value, but the flesh is considered desirable [49]. Small amounts are sold in California markets [63].

SOLE, PETRALE (Eopsetta jordani)

MEAN FAT CONTENT OF RAW OFFAL: 7.2% (6.7-7.8) [10]

HABITAT IN U.S. WATERS:
It is found in Pacific waters from California to Alaska [49].

LANDINGS IN U.S. WATERS:
Both commercial and recreational landings are made [17].

USE:
The petrale sole is considered an extremely desirable foodfish [32]. Most of the catch is sold as fillets [49].

SPOT (Leiostomus xanthurus)

MEAN FAT CONTENT OF RAW OFFAL: 10.6 [10]

HABITAT IN U.S. WATERS:
This species is very common in bay and shallow waters of the Gulf of Mexico [20]. It is also found in Atlantic waters to Cape Cod [20].

U.S. LANDINGS:
Commercial landings are made in the Gulf states [38] and in the Chesapeake Bay area [64]. The 1977 commercial harvest was 7,130,000 lbs. [15].
It is also caught by sport fishermen [38,64]. In areas off the Atlantic coast states, the sport catch is much larger than the commercial catch [64].

USE:
It is considered an excellent foodfish [25]. The commercial catch is used primarily for industrial processing into pet food [38].

SQUAWFISH (Ptychocheilus oregonensis)

MEAN FAT CONTENT OF RAW OFFAL: 10.4% [10]

HABITAT IN U.S. WATERS:
It is abundant in rivers and lakes in Oregon, Idaho, and western Montana [29].

U.S. LANDINGS:
A sport fishery exists [13].

USE:
It is used as bait [13] and consumed smoked [13].

STURGEON, ATLANTIC   (Acipenser stuvio)

MEAN FAT CONTENT OF RAW ROE: 19.7% [10]

n-3 PUFA CONTENT:
     (fillet)   (flesh)
EPA 0.11 g/100 g [26]   3.7% [56]
DHA 0.08 g/100 g [26]   2.6% [56]

HABITAT IN U.S. WATERS:
    This is an anadromous species found in coastal areas of the Atlantic and the St. Lawrence [17]. In the past, it was encountered in great numbers; presently, fish are only seen occasionally [17].

USE:
When they were plentiful, they were sought by commercial fishermen for their roe, oil and flesh [17].

STURGEON, LAKE   (Acipenser fulvenscens)

MEAN FAT CONTENT OF RAW MUSCLE: 9.1% [10]

HABITAT IN U.S. WATERS:
    It is found in lakes and large rivers west of the Appalachians, from the Great Lakes to Alabama [29]. Formerly abundant, it is now scarce in many parts of its range [13].

USE:
As a foodfish, it is often available smoked; eggs are used as caviar [17,30].

SUCKER, BUFFALO   (Ictiobus cyprinellus)

MEAN FAT CONTENT OF RAW MUSCLE: 16.6% [10]
MEAN FAT CONTENT OF RAW OFFAL: 23.1% [10]

HABITAT IN U.S. WATERS:
    It is common in the shallows of large sluggish rivers, oxbows, bayous, reservoirs and lakes in the Ohio and Mississippi river drainages, Alabama, Louisiana, Texas, Oklahoma, and Arkansas. It has been introduced in Arizona and California [29].

LANDINGS:
It is seldom caught by sport fishermen [13].

USE:
The species is an important foodfish in the Mississippi drainage region [32], and is often consumed smoked [30].

TROUT, DOLLY VARDEN (<u>Salvelinus</u> <u>malma</u>)

MEAN FAT CONTENT OF RAW MUSCLE: 6.0% (5.4-6.5%) [10]

HABITAT IN U.S. WATERS:
It is found in Washington and Alaska. Typically, it is anadromous but many populations are landlocked [29].

U.S. LANDINGS:
Recreational landings are made [36].

USE:
It is used as a foodfish [36].

TROUT, LAKE (<u>Salvelinus</u> <u>namaycush</u>)

MEAN FAT CONTENT OF RAW MUSCLE:
high fat   50.0% (33.0-66.7%) [10]
low fat   11.1% (6.9-21.8%) [10]

MEAN FAT CONTENT OF RAW OFFAL: 18.11

n-3 PUFA CONTENT:
    (fillet)
EPA 5.0 + 0.7 wt % [46]
DHA 13.4 + 1.2 wt % [46]

HABITAT IN U.S. WATERS:
It can be found in lakes in New England, in the Great Lakes region, Rocky Mountain and Pacific coast states and Alaska [29,31].

U.S. LANDINGS:
Commercial landings are made in the Great Lakes [15]. The 1977 catch totaled 356,000 lbs. [15]. Recreational landings are also made [30,31].

USE:
It is used as a foodfish [17,36].

TROUT, RAINBOW OR STEELHEAD   (Salmo gairdneri)

MEAN FAT CONTENT OF RAW MUSCLE: 6.8% (2.1-13.6%) [10]
MEAN FAT CONTENT OF RAW ROE: 9.0% (6.2-11.8%) [10]

n-3 PUFA CONTENT:
         (fillet)   (fillet)   (body oil)
EPA   5.0 wt. % [37]   0.22 g/100 g [26]   14.0 % [58]
DHA   19.0 wt. % [37] 0.62 g/100 g [26]   16.0 % [58]

HABITAT IN U.S WATERS:
This species is anadromous and has been widely introduced and established in suitable habitats in the Great Lakes region, Rocky Mountain states, Pacific coast and southwestern states [29].

U.S. LANDING:
     Commercial landings are made from Oregon to Alaska [15]. 293,000 lbs. were harvested in 1977 [15]. Recreational landings are also made [30].

USE:
It is a highly regarded foodfish [36].

TUNA, ALBACORE   (Thunnus alalunga)

MEAN FAT CONTENT OF RAW MUSCLE: 6.3% (0.7-13.2%) [10]

n-3 PUFA CONTENT:
      (white meat)   (oil)
EPA   0.63 g/100g [26]   6.5 wt % [8]
DHA   1.7 g/100g [26]   17.6 wt % [8]

HABITAT IN U.S. WATERS:
It is found in ocean waters from California to Alaska [23].

U.S LANDINGS:
     Commercial alndings are made along the Pacific coast [15]. 25,351,000 lbs. were caught in 1977 [15], 23,169,000 in 1983 [16]. Recreational landings are also made [23].

USE:
     Eight reduction plants were in existance in 1977 producing tuna and mackerel oil [15]. A total of 3,807,000 lbs. were produced in 1977 [15]; 2,535,000 lbs. in 1983 [16].

TUNA, BLUEFIN   (Thunnus thynnus)

MEAN FAT CONTENT OF RAW MUSCLE:
dark meat   8.5% (0.2-25.0%) [10]
light meat 5.0% (0.5-14.1%) [10]

Fatty Fish in U.S. Waters

n-3 PUFA CONTENT:
(fillet)
EPA 0.28 g/100g [26]
DHA 0.88 g/100g [26]

HABITAT IN U.S. WATERS:
During the summer, its distribution is in the Northeast Atlantic; in winter, it is caught over wide areas of the north and south Atlantic [65].

U.S. LANDINGS:
Commercial landings are made in California and from New Jersey to Maine [15]. In the 1970s, with a declining spawning stock due to increased fishing, the status of the Atlantic stock was considered serious. Since 1974, measures have been taken to prevent any further increase in fishing mortality [17,65]. 10,994,000 lbs. were harvested in 1977 [15]; 4,362,000 lbs. in 1983 [16]. A significant sport fishery also exists [17,47].

USE:
Oil production (see Tuna, Albacore) It is also consumed as a food fish and exported to Japan [17,65].

TUNA, SKIPJACK   (Euthynnus pelamis)

MEAN FAT CONTENT OF RAW MUSCLE: 4.5(0.2-11.0) [10]

n-3 PUFA CONTENT:
(fillet)   (raw muscle)
EPA 0.06 g/100 g [26]   10.2 % total fat [10]
DHA 0.11 g/100 g [26]   3.2 % total fat [10]

HABITAT IN U.S. WATERS:
This species can be found in Atlantic waters north and east of Cape Cod [17].

U.S. LANDINGS:
The 1977 commercial harvest was 92,203,000 lbs. [15]; 114,307,000 lbs. were caught in 1983 [16]. Recreational landings are also made [17].

USE:
Oil production (see Tuna, Albacore) It is used as a foodfish [17].

TURBOT   (Reinhardtius hippoglossoides)

MEAN FAT CONTENT OF RAW MUSCLE: 11.6% (10.8-12.4%) [10]

HABITAT IN U.S. WATERS:
The species is found in both the north Atlantic and north Pacific [45]. It is rare south of Alaska [45].

USE:
It is used as a foodfish [45].

TURBOT, ARROWTOOTH   (<u>Athresthes</u> <u>stomias</u>)

MEAN FAT CONTENT OF RAW MUSCLE: 4.8% (2.4-9.3%) [10]

HABITAT IN U.S. WATERS:
The species occurs in Pacific waters from California to Alaska [49].

USE:
It is used as mink food [45].

WHITEFISH, LAKE   (<u>Coregonus</u> <u>clupeaformis</u>)

MEAN FAT CONTENT OF RAW MUSCLE: 7.2% (1.7-16.3%) [10]
MEAN FAT CONTENT OF RAW OFFAL: 18.5% [10]

n-3 PUFA CONTENT:
                    (fillet)
EPA   6.4 wt. % [37]   0.30 g/100 g [26]
DHA   5.8 wt. % [37]   0.42 g/100 g [26]

HABITAT IN U.S WATERS:
It can be found in large rivers and lakes from the Atlantic coastal watersheds to the Northwest States and Alaska [29]. The numbers have declined recently as a result of commercial fishing and predation by the sea lamprey [31].

U.S. LANDINGS:
Commercial landings are made in the Great Lakes [15]. The 1977 harvest totaled 4,838,000 lbs [15]. Recreational landings are also made [13,30].

USE:
It is used as a foodfish [30,31]. The roe is used as caviar [30].

WHITING, KING   (<u>Menticirrhus</u> <u>saxatilis</u>)

MEAN FAT CONTENT OF RAW OFFAL: 6.0% (1.1-11.0%) [10]

HABITAT IN U.S. WATERS:
This is a rare but characteristic species found only in inshore waters from Florida to Cape Cod and in the Gulf of Mexico [20].

U.S. LANDINGS:
Commercial and sport fisheries exist [25,38].

USE:
Considered an excellent foodfish [25], it is marketed in the mid-Atlantic states [13].

REFERENCES

1. Anon. (1984). No. 1 killers: cardiovascular diseases in developed world, respiratory diseases in developing. World Health Aug/Sept: 30-31.

2. Dyerberg, J. (1981). Platelet-vessel wall interaction: influence of diet. Philos. Trans. R. Soc. Lond. B294: 373-381.

3. Dyerberg, J., and H.O. Bang (1982). A hypothesis on the development of acute myocardial infarction in Greenlanders. Scand. J. Clin. Lab. Invest. 42(161):743.

4. Hirai, A., T. Hamazaki, T. Terano, T. Nishikawa, Y. Tamura, A. Kumagai (1980). Eicosapentaenoic acid and platelet function in Japanese. Lancet ii-1132-1133.

5. Castelli, W.P. (1984). Epidemiology of coronary heart disease: the Framingham Study. Am. J. Med. 76(2A):4-12.

6. Feinleib, M. (1984). The magnitude and nature of the decrease in coronary heart disease mortality rate. Am. J. Cardiol. 54:2C-6C.

7. Tait, R.V. (1981). Elements of Marine Ecology. Butterworths & Co., London.

8. Stansby, M.E. (1976). Chemical characteristics of fish caught in the Northeast Pacific Ocean. Marine Fisheries Review Paper #1198.

9. Stansby, M.E. (1973). Polyunsaturates and fat in fish flesh. J.A.D.A. 63:625-630.

10. Sidwell, V.A. (1981). Chemical and Nutritional Composition of Finfishes, Whales, Crustaceans, Mollusks and their Products. NOAA Technical Memorandum, NMFS F/SEC-11.

11. Ackman, R.G. and C.A. Eaton (1967). Freshwater fish oils: yields and composition of oils from reduction of sheepshead, tullibee, maria and alewife. J. Fish. Res. Bd. Canada 24:1219-1227.

12. Mayo, R K. (1982). Alewife *Alosa pseudoharengus*. In *Fish Distribution*, M.D. Grosslein and T.R. Azarovitz (eds.), p.57-59. Mesa New York Bight Atlas Monograph #15. New York Sea Grant Institute, Albany.

13. McClane, A.J. (1978). *Field Guide to Freshwater Fishes of North America*. Holt, Rinehart, and Winston, New York.

14. Mayo, R.K. (1982). Blueback herring *Alosa aestivalis*. In *Fish Distribution*, M.D. Grosslein and T.R. Azarovitz (eds.), p. 54-57. Mesa New York Bight Atlas Monograph #15, New York Sea Grant Institute, Albany.

15. Fishery Statistics of the United States, 1977. *Statistical Digest* #71. National Marine Fisheries Service, 1984.

16. Fishery Statistics of the United States, 1983. *Current Fishery Statistics* #8320. National Marine Fisheries Service, 1984.

17. Thompson, P. (1980). *The Game Fishes of New England and Southwestern Canada*. Down East, Camden, Maine.

18. Richkus, W.A., and G. Dinardo (1984). *Current Status and Biological Characteristics of the Anadromous Alosid Stocks of the Eastern U.S.* Martin Marietta Environmental Systems.

19. Perlmutter, A. (1961). *Guide to Marine Fishes*. New York Univ. Press, New York.

20. Hoese, H.D., and R.H. Moore (1977). *Fishes of the Gulf of Mexico*. Texas A & M Univ. Press, College Station.

21. McClane, A.J. (1974). *Field Guide to Saltwater Fishes of North America*. Holt, RInehart and Winston, New York.

22. Hupport, D., A.D. McCall, G.D. Stauffer, K.R. Parker, J.A. McMillan, and H.W. Frey (1980). California's northern anchovy fishery: biological and economic basis for fishery management. *NOAA Tech. Mem.* NMFS SWFC-1. National Marine Fisheries Service, Washington D.C.

23. Fitch, J.E. and R.J. Lavenberg (1971). *Marine Food and Game Fishes of California*. Univ. of California Press, Berkeley.

24. Kendall, A.W. and L.P. Mercer. (1982) Black Sea Bass *Centropristis striata* In *Fish Distribution*, M.D. Grosslein and T.R. Azarovitz (eds.), p. 82-83. Mesa New York Bight Atlas Monograph #15. New York Sea Grant Institute, Albany.

25. Thomson, K.S. W.H. Weed III. A.G. Taruski, and D.E. Simanek (1978). *Saltwater Fishes of Connecticut*, 2nd edition. Connecticut Department of Environmental Protection.

26. Exler, J. and J.L. Weihrauch (1976). Comprehensive evaluation of fatty acids in foods. VIII Finfish. J.A.D.A. 69: 243-248.

27. Eldrigde, M.B., J.D. Joseph, K.M. Tabeski, and G.T. Seaborn (1983). Lipids and fatty acid composition of the endogenous energy sources of striped bass (Morone saxatilis) eggs. Lipids 18(8):510-513.

28. Smith, W.G. (1982). Striped bass Morone saxatilis In Fish Distribution, M.D. Grosslein, and T.R. Azarovitz (eds.), p. 79-82. Mesa New York Bight Atlas Monograph #15. New York Sea Grant Institute, Albany.

29. Lee, D.S., C.R. Gilbert, C.H. Hocutt, R.E. Jenkins, D.E. McAllister, and J.R. Stauffer (1980). Atlas of North American Freshwater Fishes. North Carolina State Museum of National History.

30. Eddy, S. and J.C. Underhill (1974). Northern Fishes. Univ. of Minn. Press, Minneapolis.

31. Phillips, G.L. W.D. Schmid, and J.C. Underhill (1982). Fishes of the Minnesota Region. Univ. of Minn. Press, Minneapolis.

32. Frey, H.W. (1971). California'a Living Marine Resources and their Utilization. California Dept. of Fish and Game, Sacramento.

33. Waring, G. and S. Murawaki (1982). Butterfish Peprilus triacanthus. In Fish Distribution, M.D. Grosslein and T.R. Azarovitz (eds.), p. 105-107. Mesa New York Bight Atlas Monograph #15. New York Sea Grant Institute, Albany.

34. Kimsey, J.B. and L.O. Fisk (1964). Freshwater Nongame Fishes of California. California Dept. of Fish and Game, Sacramento.

35. Worthington, R.E., T.S. Burgess Jr. and E.K. Heaton (1972). Fatty acids of channel catfish (Ictalurus punctatus). J.Fish. Res. Bd. Canada 29:113-115.

36. Wydoski, R.S. and R.R. Whitney (1979). Inland Fishes of the Minnesota Region. Univ. of Minn. Press, Minneapolis.

37. Gruger, E.H., R.W. Nelson, and M.E. Stansby (1964). Fatty acid composition of oils from 21 species of marine fish, freshwater fish, and shellfish. J. Am. Oil Chem. Soc. 41:662-667.

38. U.S. Army Corps of Engineers, Mobile District (1982). Life history requirements of selected finfish and shellfish in the Mississippi Sound and Adjacent Areas. U.S. Dept. of the Interior, Washington, D.C.

39. Gershbein, L.L. and E.J. Singh (1969). Hydrocarbons of dogfish and cod livers and herring oil. J. Amer. Oil Chem. Soc. 46:554-557.

40. Gunstone, F.D., C. Wijesunders, C.M. Scrimgeour (1978). The component acids of lipids from marine and freshwater species with special reference to furan-containing acids. J. Sci. Fd. Agric. 29:539-550.

41. Ackman, R.G., and R.D. Burgher (1964). Cod roe: component fatty acids as determined by gas-liquid chromatography. J. Fish. Res. Bd. Canada 21(3):469-476.

42. Heyerdahl, E.G. and R. Livingstone, Jr. (1982). Atlantic cod Gadus morhua. In Fish Distribution, M.D. Grosslein and T.R. Azarovitz (eds.), p. 70-72. Mesa New York Bight Atlas Monograph #15. New York Sea GRant Institute, Albany.

43. Jensen, A.C. (1972). The Cod. Thomas Y. Cromwell Co., New York.

44. Yamada, M. and K. Hayashi (1975). Fatty acid composition of lipids from 22 species of fish and mollusk. Bul. Jap. Soc. Scient. Fish. 41(11):1143-1152.

45. Eschmeyer, W.N., E.S. Herald, and H. Hammann (1983). A Field Guide to Pacific Coast Fishes of North America. Houghton Mifflin Co., Boston.

46. Kinsella, J.E., J.L. Shimp, J. Mai, and J. Weihrauch (1977). Fatty acid content and composition of freshwater finfish. J. Amer. Oil Chem. Soc. 54:424-429.

47. Clayton, G., C. Cole, and S. Murawski (1978). Common marine fishes of coastal Massachusetts. Massachusetts Cooperative Extension Service, Amherst.

48. Moriarty, C. (1978). Eels: A Natural and Unnatural History. Universe Books, New York.

49. Bane, G.W., and A.W. Bane (1971). Bay Fishes of Northern California. Mariscos Publications, Hampton Bays, New York.

50. Bell, F. H. (1981). The Pacific Halibut. Alaska Northeast Publishing Co., Anchorage.

51. Meizies, A., and I. Reichwald (1973). The lipids in flesh and roe of fresh and smoked fish. Z. Ernahrwiss 12:248-251.

52. Tocher, D.R. and J.R. Sargent (1984). Analyses of lipids and fatty acids in ripe roes of some Northwest European marine fish. Lipids 19(7);492-499.

53. Anthony, V.C. (1982). Atlantic herring Clupea harengus harengus. In Fish Distribution, M.D. Grosslein and T.R. Azarovitz (eds.), p. 61-63. Mesa New York Bight Atlas Monograph #15. New York Sea Grant Institute, Albany.

54. Berrien, P. (1982). Atlantic mackerel Scomber scombrus In Fish Distribution, M.D. Grosslein and T.R. Azarovitz (eds.), p. 99-102. Mesa New York Bight Atlas Monograph #15. New York Sea Grant Institute, Albany.

55. Reintjes, J.W. (1982). Atlantic menhaden Brevoortia tyrannus In Fish Distribution, M.D. Grosslein and T.R. Azarovitz (eds.), p. 61-63. Mesa New York Bight Atlas Monograph #15, New York Sea Grant Institute, Albany.

56. Reichwals, I., and A. Meizies (1973). Fatty acids in the flesh of freshwater and marine fish. Z. Ernahrwiss 12:86-91.

57. Glass, R.L., T.P. Krick, and A.E. Eckhardt (1974). New series of fatty acids in the northern pike (Esox lucius). Lipids 9:1004-1008.

58. Gruger, E.H. Jr. (1967). Fatty acid composition In Fish Oils: Their Chemistry, Technology, Stability, Nutritional Properties and Uses, M.E. Stansby (ed.) Avi Publishing Co., Westport, Conn.

59. Morse, W. Scup (1982) Stenotomus chrysops. In Fish Distribution, M.D. Grosslein and T.R. Azarovitz (eds.), p. 89-91. Mesa New York Bight Atlas Monograph #15. New York Sea Grant Institute, Albany.

60. Wilk, S.J. (1982). Weakfish Cynoscion regalis. In Fish Distribution, M.D. Grosslein and T.R. Azarovitz (eds.), p. 91-93. Mesa New York Atlas Monograph #15. New York Sea Grant Institute, Albany.

61. Pacheco, A.L. and L. Depres-Patanjo (1982). American Shad Alosa sapidissima In Fish Distribution, M.D. Grosslein and T.R. Azarovitz (eds.), p. 59-61. Mesa New York Bight Atlas Monograph #15. New York Sea Grant Institute, Albany.

62. Cohen, E. (1982). Spiny dogfish Squalus acanthius. In Fish Distribution, M.D. Grosslein and T.R. Azarovitz (eds.), p. 99-102. Mesa New York Bight Atlas Monograph #15. New York Sea Grant Institute, Albany.

63. Castro, J.I. (1983). The Sharks of Northern American Waters. Texas A & M Univ. Press, College Station.

64. Selverman, M.J. (1982). Spot Leiostomus xanthurus. In Fish Distribution, M.D. Grosslein, and T.R. Azarovitz (eds.) p. 93-95. Mesa New York Bight Atlas Monograph #15. New York Sea Grant Institute, Albany.

65. Schuck, H.A. (1982). BLuefin Tuna Thunnus thynnus. In Fish Distribution, M.D. Grosslein and T.R. Azarovitz (eds.), p. 102-105. Mesa New York Bight Atlas Monograph #15. New York Sea Grant Institute, Albany.

66. Exler, J. (1985). U.S.D.A. Nutrient Record Data Base.

67. Exler, J. and J.L. Weihrauch (1985). Provisional table on the content of omega-3 fatty acids and other fat components of selected foods. U.S.D.A. Human Nutrition Information Service, HNIS/PT-103.

68. Bruckner, G.G., B. Lokesh, B. German, and J.E. Kinsella (1984). Biosynthesis of prostanoids, tissue fatty acid composition, and thrombotic parameters in rats fed diets enriched with docosahexaenoic or eicosapentaenoic acids. Thromb. Res. 34:479-497.

Index

ADP 83, 129
    Platelets 129

Adipose 155
    Yellow 155

Adrenalin

Aggregation 9, 83
    ADP 83
    Adrenaline 83
    Cod Liver Oil 84
    Collagen 83
    Fish oils 83-85
    PGH2 84
    Platelets 9, 83
    Salmon 84
    Thrombin 84

Alewife 262
    Habitat 262
    Landings 262
    Uses 262

Amberjack - Greater 263
    Habitat 263
    Landings 263
    Uses 263

Animals 107
    Feeding trials 107-160
    Subjects 107

Anchovy - European 241, 263
    Cholesterol 241
    Fatty Acids 241
    N-3 PUFA Content 241
    Habitat 263
    Landings 263
    Proximate Composition 241
    Uses 263

Antithrombin III 78
    Fish oil 78

Aorta
    Contraction 140
    Diet 138
    Lipids 138, 140
    Norepinephrine 141
    Prostacyclin synthesis 138-139

Arachidonic Acid 27 see polyun-
    saturated
    Acylation 27
    Aorta 152
    Exogenous 139
    Fish 241-259
    Fish oil 68
    Genetics 100
    Leukotrienes 33
    Lung 152
    Metabolism 29
    Platelet 79

Arterial Tissue 2
    Contraction 140
    DHA 140
    EPA 140
    Fish oil 140
    Thrombosis 140
    Wall, changes 2

Arthritis 234

Asthma 232
    Leukotrienes 232

Atherosclerosis 1, 80
    Erythrocytes 94
    Platelet number 83
    Prostacyclin 91
    Thromboxane 86

Autoxidized Fish Oils 185
    Antioxidants 196
    Feeding 195
    Rats 195
    Safety 185

Autoxidation 193
    Fish oils 194
    In vivo 196
    Linolenic acid 194
    Vitamin E 196

Bass - Black Sea 263
    Habitat 264
    Landings 264
    Uses 264

Bass - Largemouth 241
    Cholesterol 241
    Fatty Acids 241
    N-3 PUFA Content 241
    Proximate Composition 241

Bass - Rock 241
    Cholesterol 241
    Fatty Acids 241
    N-3 PUFA Content 241
    Proximate Composition 241

Bass - Stripped 241, 264
    Cholesterol 241
    Fatty Acids 241
    N-3 PUFA Content 245
    Habitat 264
    Landings 264
    Proximate Composition 241
    Uses 264

Bass - White 242
    Cholesterol 242
    Fatty Acids 242
    N-3 PUFA Content 242
    Proximate Composition 242

Bleeding Tendency 18

Bleeding Time 87, 94, 132
    Fish oils 132

Bloater 264
    Habitat 264
    Landings 264
    Uses 264

Blood
    Pressure 4, 95, 141
    Viscosity 94-95

Blood Pressure 4,
    Fish oil 141

Bonito - Atlantic 264
    Habitat 265
    Landings 265
    Uses 265

Bonito - Pacific 265
    Habitat 265
    Landings 265
    Uses 265

Brain 156, 157
    Blood flow 158
    Ischemia 157
    Neurological 156
    Occlusion 158
    Thromboxane 157

Bullhead - Black 265
    Habitat 265
    Landings 265
    Uses 265

Butterfish 265
    Habitat 266
    Landings 266
    Uses 266

Burbot 242
    Cholesterol 242
    Fatty Acids 242
    N-3 PUFA Content 242
    Proximate Composition 242

Cadmium 203

Cancer 34, 165-175
    Breast 166
    Leukotrienes 34
    Mammary 166
    Metastasis 165
    n-6 PUFA 165
    PGE2 166
    Prostaglandins 167
    Tumor 165

Capelin 242
    Cholesterol 242
    Fatty Acids 242
    N-3 PUFA Content 242
    Proximate Composition 242

Index

Cardiac see heart
    Erucic acid 198
    Infarction 138
    Lipidosis 198
    Lipids 133-136

Cardiovascular 12
    Disease 1, 160, 161
    Eskimos 18
    Mortality 12
    Polyunsaturated fatty acids 6
    Risk factors 3, 12

Carnitine 169
    Palmitoyltransferase 149

Carp 242, 266
    Cholesterol 242
    Fatty Acids 242
    N-3 PUFA Content 242
    Habitat 266
    Landings 266
    Proximate Composition 242
    Uses 266

Catfish - Brown Bullhead 243
    Cholesterol 243
    Fatty Acids 243
    N-3 PUFA Content 243
    Proximate Composition 243

Catfish - Channel 243, 266
    Cholesterol 243
    Fatty Acids 243
    N-3 PUFA Content 243
    Habitat 266
    Landings 266
    Proximate Composition 243
    Uses 266

Cerebral see brain

Chemotaxis 34

Cholesterol 4, 74-77, 146
    Decrease mechanisms 6
    Dietary 4
    DHA 159
    EPA 159
    Esters 123
    Fish 177-180, 241-259
    HDL 123, 159
    Intake 12
    LDL 123
    Liver 146
    Lung 152
    n-3 PUFA 159
    Plasma 4, 74-77
    Rat plasma 122-123
    VLDL 123

Chub - Utah 267
    Habitat 267
    Landings 267
    Uses 267

Cisco - Lake 267
    Habitat 267
    Landings 267
    Uses 267

Cisco - Longjaw 268
    Habitat 268

Cisco - Shortjaw 262
    Habitat 262
    Uses 262

Cisco - Shortnose 262
    Habitat 262
    Uses 262

Clotting
    Thromboxane 85

Clupanodonic acid
    Fish 241-258
    Fish oil 259, 26

Cobia 262
    Habitat 262
    Landings 262
    Uses 269

Coconut Oil
    Hydrogenated 152

Cod - Atlantic 243, 269
    Cholesterol 243
    Fatty Acids 243
    N-3 PUFA Content 243, 269
    Habitat 269
    Landings 269
    Proximate Composition 243
    Uses 269

Cod - Pacific 243, 269
    Cholesterol 243
    Fatty Acids 243
    N-3 PUFA Content 243, 269
    Habitat 269
    Landings 269

Proximate Composition 243
Uses 269

Cod Liver Oil 44, 45, 66, 91, 260
    Anti-thrombin 78
    Arachidonic acid 68
    Bleeding time 87
    Blood pressure 95
    Composition 66, 114, 260
    Cholesterol 260
    Fatty Acids 260
    Erythrocytes 92
    Kidney 153
    Malondialdehyde 84
    Max EPA 78
    N-3 PUFA Content 260
    Plasma characteristics 50
    Platelet aggregation 59, 84
    Prostacyclin 90
    Thrombin 84
    Thromboxane B3 87
    Thromboxane 91

Collagen 3
    Cod liver oil 84
    Platelet aggregation 83, 130, 132
    Thromboxane 85

Composition
    of fish 241-259
    of fish oil 260-261

Consumption
    Fish 171-175
    Seafood 171

Corn oil 146, 154 see feeding
    Adipose 156
    Kidney 154

Croaker - Atlantic 244
    Cholesterol 244
    Fatty Acids 244
    N-3 PUFA Content 244
    Proximate Composition 244
    Uses 244

Cyclooxygenase 26
    Fish oil 30

Danish 14
    Diet 14, 16
    Dietary fatty acids 16
    Lipids 14
    Lipoproteins 14, 15
    Plasma lipids 15
    Platelets 14,15

Desaturase 147
    Liver 147
    Fish oil 147, 149

Desaturation 8

Diet 4, 14
    Eskimos 14
    Fat 4
    Polyunsaturated 5
    Saturated fatty acids 5

Dietary Fat 8 see feeding
    Danes 16
    Eskimos 16
    Leukotrienes 35
    Platelets 7

Dietary Studies - see feeding trials
    Cholesterol 4, 5
    Blood 62
    Blood pressure 62-66
    Fish 42-56
    Fish oils 42-56
    Humans 41-100
    Human subjects 65
    Max EPA 48-49
    Plasma lipids 50-55
    Platelets 56
    Platelet aggregations 57, 58
    Research needs 235

Docosahexaenoic Acid 241-294
    Brain 156
    Consumption 172, 173
    Dane 17
    Eskimo 17
    Fish 241-294
    Fish content 171-173, 241-259
    Heart 136
    Lipoxygenase 36
    Liver 145, 146
    Lung 151
    Plasma cholesterol 122, 123
    Platelet 82, 129
    Tissue 159

Dogfish - Spiny 244
    Cholesterol 244
    Fatty Acids 244
    N-3 PUFA Content 244
    Proximate Composition 244

Dolphin fish 244
    Cholesterol 244
    Fatty Acids 244
    N-3 PUFA Content 244

Index

Drum - Black 244
    Cholesterol 244
    Fatty Acids 244
    N-3 PUFA Content 244
    Proximate Composition 244

Drum - Freshwater 254, 270
    Cholesterol 245
    Fatty Acids 245
    N-3 PUFA Content 270
    Habitat 270
    Landings 270
    Proximate Composition 245
    Uses 270

Eel, Common 270
    N-3 PUFA Content 270
    Habitat 270
    Landings 270
    Uses 270

Eel, Conger 271
    Landings 271
    Uses 271

Eicosapentaenoic Acid, EPA 6,7, 65, 241-294
    Aorta 138
    Brain 156, 157
    Consumption 172, 173
    Conversion 139
    Dane 17
    Dietary changes 70
    EPA 6
    Eskimo 17
    Esters 138
    Feeding trials 48, 49, 65, 124
    Fish 241-259
    Fish content 171
    Heart 126
    Leukotrienes 35
    Lipoxygenase 36, 132
    Liver 145
    Lung 151, 152
    Neutrophils 35
    Platelet 79-80, 81, 82, 129,131
    Tissue 159

Eicosanoids 9, 25
    Antagonism 30
    Arthritis 234
    Functions 31
    Production 26
    Synthesis 28
    Thromboxane 9

Endothelial 2

Enzymes 7
    Desaturases 8
    Desaturation 8

EPA, 65 see Eicosapentaenoic acid
    Blood pressure 95
    Dietary 65
    Ingestion 65
    Phospholipids 61
    Plasma cholesterol 122-123
    Platelets 61, 79-84
    Prostacyclin 90-93
    Thromboxane 85

Epidemiology 19, 41
    Fish consumption 19
    Fish oil 41
    Japanese 19

Erucic acid 197
    Fish 241-259
    Lipidosis 197
    Rat 198

Erythrocyte
    DHA 92, 93
    EPA 92, 93
    Fatty acids 92
    Fish 92
    Fish oils 92
    Phospholipids 92

Essential Fatty Acids 9, 10, 160
    Deficiency 160
    Linoleic 10
    n-3 PUFA 11

Erythrocytes
    Fish oil 64, 92-94
    Phospholipids 64-66

Eskimos 13
    Diet 14, 16
    Dietary fatty acids 16
    Diseases 13
    Lipids 15
    Lipoproteins 14
    Phospholipids 14
    Plasma 15
    Platelets 17
    Thromboxane 17

Eulachon 271
    Habitat 271
    Landings 271
    Uses 271

Excretion
    Prostacyclin metabolites 91, 92

Experimental Design 96

Fatty Fish 239, 262-293
    Bleeding time 87, 89
    Feeding trials 42-90
    HDL - cholesterol 77
    Herring 43, 44
    LDL - cholesterol 77
    Mackerel 43
    Plasma characteristics 50
    Plasma fatty acids 70
    Platelets 56-59
    Prostacyclin 90
    Salmon 42, 43, 44
    VLDL - cholesterol 76, 77

Fatty Acids
    of Fish 241-261
        Anchovy - European 241
        Bass - Black Sea
        Bass - Largemouth 241
        Bass - Rock 241
        Bass - Stripped 241
        Bass - White 242
        Burbot 242
        Capelin 242
        Carp 242
        Catfish - Brown Bullhead 243
        Catfish - Channel 243
        Cod - Atlantic 243
        Cod - Pacific 243
        Cod liver 260
        Croaker - Atlantic 244
        Dogfish - Spiny 244
        Dolphin fish 244
        Drum - Black 244
        Drum - Freshwater 245
        Flounder 241
        Flounder - Yellowtail 245
        Grouper - Red 245
        Grouper - Jewfish 246
        Haddock 246
        Hake - Atlantic 246
        Hake - Pacific 246
        Hake - Red 247
        Hake - Silver 247
        Hake - Squirrel 247
        Halibut - Atlantic 247
        Halibut - Pacific 247
        Halibut - Greenland 248
        Herring - Atlantic 248
        Herring - Pacific 248

Mackerel - Atlantic 248
Mackerel - Chub 249
Mackerel - Horse 249
Mackerel - Japanese Horse 249
MaxEPA 260
Menhaden - Atlantic 249
Mullet - Striped 250
Perch - Ocean 250
Perch - Yellow 250
Perch - White 250
Pike - Northern 251
Pike - Walleye 251
Plaice - European 251
Pollock 251
Pompano - Florida 252
Rockfish - Canary 251
Sablefish 252
Salmon - Atlantic 252
Salmon - Chinook 253
Salmon - Chum 253
Salmon - Coho 253
Salmon - Pink 253
Salmon - Sockeye 254
Sandlance - American 254
Seatrout - Sand 254
Seatrout - Spotted 254
Shark 260
Shark - Grayfish 287
Smelt - Rainbow 257
Snapper - Red 255
Sole - European 255
Sprat 255
Sucker - White 256
Sunfish - Pumpkinseed 256
Swordfish 256
Trout - Artic Char 257
Trout - Brook 257
Trout - Lake 257
Trout - Rainbow 256
Tuna - 257
Tuna - Albacore 258
Tuna - Bluefin 258
Tuna - Skipjack 158
Tuna - Yellowfin 158
Whitefish - Lake 259
Wolffish - Atlantic 259
Whiting 259

Feeding Trials 42-56
    Animals 107-160
    Atherosclerosis 80, 94
    Bleeding time 87, 88
    Blood pressure 95
    Brain 156
    Cancer 166
    Cats 108-112

Index

Clotting 85
Coconut oil 152
Cod liver oil 41, 44
Design 96, 97
Docosahexaenoic 42-48
Dogs 108-112
Duration 81, 99
Eicosapentaenoic 42, 48
EPA 48, 49, 70
Erythrocytes 92-94
Eskimos 79
Fatty fish 42
Fish oil 42-48
Gerbils 108-112
Guinea pigs 108-112
HDL - cholesterol 76-77
Herring 43, 44
Human 41-100
Kidney 153, 155
LDL and VLDL cholesterol
Lung 150
Mackerel 42, 44
Marine oils 107-190
Max- EPA 43, 46-48
n-3 PUFA consumption 174
Plasma changes 68
Plasma cholesterol 74-80
Plasma fatty acids 68
Plasma triglycerides
Platelet aggregation 83
Platelet number 83
Platelets 79-87
Prostacyclin 90-92
Rat liver 148
Research needs 235
Reversibility 81
Salmon 41, 42, 44
Salmon oil 41
Subjects 65
Thromboxane 85-87

Fish 41
   Ash 241-258
   Catch 209, 261-294
   Cholesterol 177, 182, 241-262
   Composition 241-261
   Consumption 19, 172
   Fat 241-261, 261-294
   Fatty 240
   Fatty acids 241-263 see fatty acids
   Heavy metals 203
   Lean 171, 173
   Lipid content 239, 240-294
   Mackerel 41
   Meal 212
   Nutrients 241-261
   PCB 200
   Pesticides 199-200
   Platelet aggregation 84
   Prostacyclin 90
   Processing 209-220
   Proximate composition 241
   Salmon 41
   Sources 239
   Tocopherol 188
   Vitamin A 182-184
   Vitamin D 185
   Vitamin E 186

Fish Oils 41, 12
   AA metabolism 160
   Anti-inflammatory effects 36
   Antioxidant 220
   Autoxidation 193
   Autoxidized 195
   Blood pressure 95
   Brain 156
   Cancer 166, 167
   Cholesterol 177, 250-260
   Cod liver oil 44, 45
   Composition 66, 67, 114-115, 211, 260-261
   Consumption 136, 137, 171-175
   Control Standards 223
   Demethylation 149
   Deodorization 217
   Desaturase 147
   Dosage 100
   Exports 212
   Fatty acids 260
   Feeding 71, 124
   Fractionation 222
   HDL - cholesterol 16
   Health effects 231
   Hydrogenation 217
   Hypolipidemic 231
   Ischemia 157
   6-keto F1alpha excretion 91
   Kidney 155
   Leukotriene B4 36
   Leukotriene C4 36
   Liver 146, 147
   Liver enzymes 147
   Lung 150
   Max-EPA 43, 46, 47, 48
   Membrane 149
   Menhaden 260
   Metabolic effects 155-160
   Metals 203
   n-3 PUFA 171-174
   Optimum dosage 236

Oxidative 149
Packaging 225
Pesticides 199, 201
Pesticide removal 219
Plasma EPA, DHA 114
Platelets 79, 87
Platelet aggregation 57-59, 83
Processing 209, 213-219
Production 210
Prostacyclin 139, 160
Proximate composition 260, 261
Quality control 223
Rats 136
Reduce AA metabolism 36
Research needs
Refining 215, 216
Retinol 182
Safety 193, 198
Sardine oil 46
Side effects 97
Species used 210
Stability 219
Storage 224
Supply 213
Technology 209
Thromboxane 160
Tocopherol 187, 220
Tumors 166-170
Vitamin A 179-182
Vitamin D 185
Vitamin E 185

Fish Oils Effects 31
    Cardiac functions 133
    Cardiac lipids 136
    Composition 114, 115
    Platelet fatty acids 80-82
    Plasma lipids 113
    Prostaglandins 31
    Reversibility 81
    Thromboxane 85, 131

Flounder 241, 271
    Cholesterol 245
    Fatty Acids 245
    N-3 PUFA Content 245
    Habitat 271
    Landings 271
    Proximate Composition 245
    Uses 271

Flounder - Yellowtail 245
    Cholesterol 245
    Fatty Acids 245
    N-3 PUFA Content 245
    Proximate Composition 245

Gadoleic Acid
    Fish 241-260

Gerbils 156-157
    Ischemia 157
    Menhaden feeding 156

Grouper - Red 245
    Cholesterol 245
    Fatty Acids 245
    N-3 PUFA Content 245
    Proximate Composition 245

Grouper - Jewfish 246
    Cholesterol 246
    Fatty Acids 246
    N-3 PUFA Content 246
    Proximate Composition 245

Guinea Pig
    Fish oil 151
    Lung lipids 151
    Prostaglandins 152

Haddock 246
    Cholesterol 246
    Fatty Acids 246
    N-3 PUFA Content 246
    Proximate Composition 246

Hake - Atlantic 246
    Cholesterol 246
    Fatty Acids 246
    N-3 PUFA Content 246
    Proximate Composition 246

Hake - Pacific 246
    Cholesterol 246
    Fatty Acids 246
    N-3 PUFA Content 246
    Proximate Composition 246

Hake - Red 247
    Cholesterol 247
    Fatty Acids 247
    N-3 PUFA Content 247
    Proximate Composition 247

Hake - Silver 247
    Cholesterol 247
    Fatty Acids 247
    N-3 PUFA Content 247
    Proximate Composition 247

Hake - Squirrel 271
    Habitat 272

Index

Landings 272
Uses 272

Halibut - Atlantic 247, 272
    Cholesterol 247
    Fatty Acids 247
    N-3 PUFA Content 247, 272
    Habitat 272
    Landings 272
    Proximate Composition 247
    Uses 272

Halibut - Pacific 247, 272
    Cholesterol 247
    Fatty Acids 247
    N-3 PUFA Content 247, 272
    Habitat 272
    Landings 272
    Proximate Composition 247
    Uses 272

Halibut - Greenland 248
    Cholesterol 248
    Fatty Acids 248
    N-3 PUFA Content 248
    Proximate Composition 248

HDL - Cholesterol 76-77

Heart Disease 17
    in Eskimos 17

Hematocrit 62-66, 94

Herring - Atlantic 248, 273
    Cholesterol 248
    Fatty Acids 248
    N-3 PUFA Content 248, 273
    Habitat 273
    Landings 273
    Proximate Composition 248
    Uses 273

Herring - Pacific 248, 273
    Cholesterol 248
    Fatty Acids 248
    N-3 PUFA Content 248
    Habitat 273
    Landings 273
    Proximate Composition 248
    Uses 273

Herring - Round 274
    Habitat 274

Herring
    Blood pressure 95
    DDT 202
    Feeding 44
    Oil composition 114, 260
    PCB 202

Herring Oil 260
    Cholesterol 260
    Human trials 41-100

Hydroxy Fatty Acids 132
    HETE 132, 133
    HHT 132, 133
    Lipoxygenase

Hypertension 2

Hypertriglyceridemia
    Dietary fish oil 77

Infarction 3
    Cardiac
    Cerebral 3

Inflammation 32
    Leukotrienes 34

Ischemia 157, 158
    Brain 157

Japanese 19

Kidney 153, 154
    Fish oil 153, 154
    Lipids 153
    Prostaglandin 153

Kiyi - 274
    Habitat 274
    Uses 274

Lamprey, River 274
    Habitat 274

LDL - Cholesterol 77, 78
    Apoprotein synthesis 77
    Salmon oil 77

Leukemic Cells 167
    Chemotherapeutic 167

Leukotrienes 32
    Actions 34
    Asthma 232
    Chemotaxis 35
    EPA 35
    Fish oil 36
    Functions 32, 34

Inflammation 32, 34
Inflammatory 234
Leukotriene B4 33
Leukotriene C4 33
Menhaden oil 35
Platelet 235
Psoriasis 234
PUFA 35
Synthesis 32, 33

Lipogenesis 159

Lingcod 274
Habitat 274
Landings 274
Uses 274

Linoleic acid 6,7
Acylation 27
Consumption 10
Effects of EPA & DHA 72
Fish oil effects 12
Metabolism 11, 27, 28
Platelets 79, 80
Requirement 10

Lipids
Liver 145, 146
Peroxide 126, 150

Lipoproteins 2, 14
Reduction 5

Lipoxygenase 26
EPA 36
DHA 36
Leukotrienes 32, 33

Liver 144
Cholesterol 146
Enzymes 148
Fatty acids 145-146
Lipids 145
Pig 145
Triglycerides 146
Weight 144

Lung
Fatty acids 151, 152
Fish oil 150
Lipids 150, 151
Prostaglandin 151
Thromboxane 151

Mackerel
Feeding trial 42

Prostacyclin I3 90
Thromboxane 85

Mackerel - Atlantic 248, 275
Cholesterol 248
Fatty Acids 248
N-3 PUFA Content 248, 275
Habitat 275
Landings 275
Proximate Composition 248
Uses 275

Mackerel - Chub 249, 275
Cholesterol 249
Fatty Acids 249
N-3 PUFA Content 249, 275

Mackerel - King 276
Habitat 276
Landings 276

Mackerel - Horse 249
Cholesterol 249
Fatty Acids 249
N-3 PUFA Content 249
Proximate Composition 249

Mackerel - Japanese Horse 249
Cholesterol 249
Fatty Acids 249
N-3 PUFA Content 249
Proximate Composition 249

Mackerel - Spanish 276
Habitat 276
Landings 276
Uses 276

Macrophages 3, 167-170
Cancer 167
n-3 PUFA 167
PGE 2 167
Tumor 167

Malondialdehyde 89, 132

Marine Oils 146
Cholesterol 180
Effects 159
Enzymes 147
Feeding 150, 159
Liver 146
Pesticides 200

Mass-Spectrometry 91, 153, 154
Prostacyclin 91

Index

Max EPA 46-47, 48, 65, 66, 260
    Animal trials 107-130
    Atherosclerosis 80, 83, 86, 91
    Bleeding time 89
    Blood pressure 95
    Blood viscosity 94-95
    Cholesterol 75, 260
    Composition 260
    DHA 80-83, 260
    Dietary 46
    EPA platelet 80-83
    Fatty acids 260
    Feeding 46
    PGE2 effects 84
    Plasma cholesterol 122
    Plasma fatty acids 68
    Plasma triglycerides 71, 72, 73, 78
    Platelets 55, 60, 79-83
    Platelet aggregation 83
    Platelet count 83
    Prostacyclin 91
    Rabbit 130
    Thromboxane 85, 86, 91

MaxEPA 260
    Cholesterol 260
    Fatty Acids 260
    N-3 PUFA Content 260

Membrane Fluidity 149

Menhaden - Atlantic 249
    Cholesterol 249
    Fatty Acids 249
    N-3 Content 249, 276
    Habitat 276
    Landings 276
    Proximate Composition 249
    Uses 276

Menhaden - Gulf 277
    Habitat 277
    Landings 277
    Uses 277

Menhaden Oil 260
    Bleeding time 132
    Blood pressure 141
    Brain 156, 157-159
    Cancer 166, 168
    Cholesterol 260
    Composition 114, 218, 260
    Enzymes 147
    Heart lipids 137-139
    Kidney 153, 155
    Lung 151
    PGE2 166
    Plasma 113
    Rats
    Thromboxane 131
    Plasma 113
    Tumor 166

Mercury 203-204

Mullet - Striped 250, 277
    Cholesterol 250
    Fatty Acids 250
    N-3 PUFA Content 250, 277
    Habitat 278
    Landings 278
    Proximate Composition 250
    Uses 278

Neurological - see brain

Neutrophils 35
    EPA 35, 36
    Fish oil 36
    Human 36
    Lipoxygenase 36

Norepinephrine 141

Nucleotidase 149

Oleyl CoA 169

Pathophysiological Effects 34
    Fish oils 36, 37
    Leukotrienes 34
    Psoriasis 34

Perch - Ocean 250, 278
    Cholesterol 250
    Fatty Acids 250
    N-3 PUFA Content 250, 278
    Habitat 278
    Landings 278
    Proximate Composition 250
    Uses 278

Perch - Yellow 250, 279
    Cholesterol 250
    Fatty Acids 250
    N-3 PUFA Content 250, 279
    Habitat 279
    Landings 279
    Proximate Composition 250
    Uses 279

Perch - White 250
    Cholesterol 250
    Fatty Acids 250
    N-3 PUFA Content 250
    Proximate Composition 250

Perch - Pacific Ocean 278
    N-3 PUFA Content 279
    Habitat 279
    Landings 279
    Uses 279

Perch, Blue Walleye 278

Peroxides 126, 160, 189, 195
    Value 195

Pesticides 189
    DDT 189
    Fish 200
    Fish oil 201-202
    PCB 200
    Polychlorinated biphenyls 189

PGE2 167
    Macrophages 167

Phosphatidylcholine 81-82
    Cardiac 139
    Erythrocytes 92-94

Phosphatidylinositol 82
    Cardiac 139
    Erythrocytes 92-94

Phosphatidylserine 82
    Cardiac 137
    Erythrocytes 92-94
    n-3 PUFA 137

Phospholipids 56
    Cardiac 136
    Erythrocytes 92
    Fish effects 56
    Microsomes 151
    Phosphatidylcholine 57
    Plasma 69
    Platelets 56

Pigs 155
    Adipose 155
    Erucic acid 198

Pike - Northern 251, 179
    Cholesterol 251
    Fatty Acids 251

N-3 PUFA Content 251, 279
    Habitat 280
    Landings 280
    Proximate Composition 251
    Uses 280

Pike - Walleye 251
    Cholesterol 251
    Fatty Acids 251
    N-3 Content 251
    Proximate Composition 251

Pilchard 280
    N-3 PUFA Content 280
    Habitat 280
    Landings 280
    Uses 280

Plaice - European 251
    Cholesterol 251
    Fatty Acids 251
    N-3 PUFA Content 251
    Proximate Composition 251

Plasma
    Animal 113
    Cholesterol 74-77
    Fatty acids 68, 113
    Feeding trial 50-55, 113
    Fish oil 50-55
    Lipids 50-55, 113-115
    Triglycerides 71, 74, 120-121

Plasma Characteristics 50-55, 68
    Lipids 51-55
    Lipoproteins 51, 55

Plasma Cholesterol 74-77
    EPA 75
    Fish oils 74-77
    HDL - cholesterol 76, 77, 123
    LDL & VLDL cholesterol 177, 123
    Max EPA 75

Plasma Lipids
    Rat 113

Plasma Triglycerides 70-71
    Dietary effects 70-74
    Max EPA 78

Platelets 2
    Adherence 235
    Aggregation 9, 58, 129, 130
    Arachidonic acid 79
    Collagen 130

Index

DHA 79, 126
Dietary fat 8, 56, 126
EPA 79, 126
Eskimo 17
Fatty acids 79, 125, 126
Fatty fish 56
Fish oils 56-60, 79, 87
Leukotrienes 235
Linolenic acid 79
Number 59, 129
Phosphatidylcholine 81, 82
Phosphatidylserine 82
Phosphatidylinositol 82
Phospholipids 56, 80, 81, 82, 126-128
Platelets 83
Platelet aggregation 83, 129
Reversibility of change 81, 82
Thromboxane 85
Vessel wall interaction

Platelet Aggregation
    Thromboxane 85-87

Pollock 251, 280
    Cholesterol 251
    Fatty Acids 251
    N-3 PUFA Content 251, 280
    Habitat 280
    Landings 280
    Proximate Composition 251
    Uses 281

Polyunsaturated Fatty Acids N-3 - 5, 41, 231
    Autoxidation 197
    Cancer 167-170
    Consumption 171-176
    Content 172, 173
    Dietary effects 41-100
    Macrophages 167
    Mechanisms 233
    Optimum intake 236

Polyunsaturated Fatty Acids (PUFA) 5, 10
    Cancer 165
    Cardiac 133, 137, 139
    Cardiovascular disease 6
    Cholesterol 5
    Dietary 6
    Dosage
    Leukotrienes 35
    Lipoproteins 5
    Mechanisms 6

    Metabolic interrelationships 8, 11
    n-3/n-6 interactions 98
    Peroxides 150
    Pools 99
    Prostaglandin 152
    P/S ratios 5

Polyunsaturated Fatty Acids N-6 5
    Cancer 165

Pompano - Florida 252
    Cholesterol 252
    Fatty Acids 252
    N-3 PUFA Content 252
    Proximate Composition 252

Prostacyclin 9
    Antiaggregatory 9, 138
    Aorta 138
    Diet 90
    Effects 30
    Eskimo 17
    Fish 90
    Fish oil 31
    Prostacyclin I3, 90
    Rats 139
    Vasodilation 9

Prostaglandins 26, 90
    Antagonism 30
    Cancer 166
    Formation 152
    Functions 26, 30, 31
    Kidney 153, 154
    Immunological effects 166
    Lung 152, 153
    Macrophages 167
    n-3 PUFA 167
    PGE2 cancer 166-170
    Synthesis 28, 29

PUFA see polyunsaturated fatty acids

Rabbit 156
    Max EPA 156

Radioimmunoassay 153

Rat
    Autoxidized fish oil 185
    Blood pressure 141
    Cancer 166
    EPA: DHA ratio 127
    Fish oil 136, 137
    Liver desaturase 148
    n-3 PUFA 119, 128

Plasma fatty acids 117, 120
Plasma cholesterol 122
Plasma triglycerides 120, 121
Platelets 129
Prostacyclin 138-139
Thromboxane 131
Tumor 166

Ratfish 281
    Habitat 281
    Uses 281

Research 235
    Metabolic pools 98
    Needs 98, 99, 235, 236

Retinol 183

Risk Factors 3

Rockfish - Canary 251
    Cholesterol 251
    Fatty Acids 251
    N-3 PUFA Content 251
    Proximate Composition 251

Sablefish 252, 281
    Cholesterol 252
    Fatty Acids 252
    N-3 PUFA Content 252, 281
    Habitat 281
    Landings 281
    Proximate Composition 251

Safflower Oil 149

Salmon 72
    Malondialdehyde 89
    Thrombin 84

Salmon Oil 261
    Cholesterol 261
    Composition 261

Salmon - Atlantic 252, 282
    Cholesterol 252
    Fatty Acids 252
    N-3 PUFA Content 252, 282
    Habitat 282
    Landings 282
    Proximate Composition 252
    Uses 282

Salmon - Chinook 253, 282
    Cholesterol 253
    Fatty Acids 253

N-3 PUFA Content 253, 282
Habitat 282
Landings 282
Proximate Composition 253
Uses 282

Salmon - Chum 253, 283
    Cholesterol 253
    Fatty Acids 253
    N-3 PUFA Content 253, 283
    Habitat 283
    Landings 283
    Proximate Composition 253
    Uses 283

Salmon - Coho 253, 284
    Cholesterol 253
    Fatty Acids 253
    N-3 Content 253, 284
    Habitat 284
    Landings 284
    Proximate Composition 253
    Uses 284

Salmon - Pink 253, 283
    Cholesterol 253
    Fatty Acids 253
    N-3 PUFA Content 253, 283
    Habitat 283
    Landings 283
    Proximate Composition 253
    Uses 283

Salmon - Sockeye 254, 284
    Cholesterol 254
    Fatty Acids 254
    N-3 PUFA Content 254, 284
    Habitat 284
    Landings 284
    Proximate Composition 253
    Uses 283

Sandlance - American 254
    Cholesterol 254
    Fatty Acids 254
    N-3 PUFA Content 254
    Proximate Composition 254

Sardine Oil 45
    Arachidonic acid 68
    Blood pressure 95
    Composition 67, 114
    Dietary 45
    Feeding 45
    Plasma FA 68, 69, 113
    Platelets 59

Index

Thromboxane 85

Saury - Atlantic 285
    Habitat 285
    Landings 285
    Uses 285

Saury - Pacific 285
    Habitat 285
    Landings 285
    Uses 285

Scad 285

Scup 285
    Habitat 286
    Landings 286
    Uses 286

Seafood
    Consumption 172

Seatrout - Gray 286
    Habitat 286
    Landings 286
    Uses 286

Seatrout - Sand 254
    Cholesterol 254
    Fatty Acids 254
    N-3 Content 254
    Proximate Composition 254

Seatrout - Spotted 254
    Cholesterol 254
    Fatty Acids 254
    N-3 Content 254
    Proximate Composition 254

Shad 286
    Habitat 286
    Landings 286
    Uses 286

Shad - Gizzard 287
    Habitat 287
    Landings 287
    Uses 287

Shad - Hickory 287
    Habitat 287
    Landings 287
    Uses 287

Shark 151
    Liver oil 151, 260

Shark Oil 260, 261
    Cholesterol 260
    Fatty Acids 261

Shark - Grayfish 287
    N-3 PUFA Content 287
    Habitat 287
    Landings 287
    Uses 287

Shark - Greenland 288

Shark - Salmon 289
    N-3 Content 288
    Habitat 288
    Landings 288

Shark - Sevengill 288
    Habitat 288
    Uses 288

6-keto Flalpha 129

Smooth Muscle 2, 3

Smelt - Rainbow 255, see Eulachon
    Cholesterol 255
    Fatty Acids 255
    N-3 PUFA Content 255
    Proximate Composition 255

Snapper - Red 255
    Cholesterol 255
    Fatty Acids 255
    N-3 PUFA Content 255
    Proximate Composition 255

Sole - European 255
    Cholesterol 255
    Fatty Acids 255
    N-3 PUFA Content 255
    Proximate Composition 255

Sole - Petrale 289
    Habitat 289
    Landings 289
    Uses 289

Sprat
    Cholesterol 255
    Fatty Acids 255
    N-3 Content 255
    Proximate Composition 255

Spot - 289
    Habitat 289

Landings 289
Uses 289

Squawfish 289
    Habitat 289
    Landings 289
    Uses 289

Squid 144
    Liver Composition 114
    Peroxide 150

Sturgeon - Atlantic 290
    N-3 PUFA Content 290
    Habitat 290
    Landings 290
    Uses 290

Sturgeon - Lake 290
    Habitat 290
    Landings 290
    Uses 290

Sucker - Buffalo 290
    Habitat 290
    Landings 290
    Uses 290

Sucker - White 256
    Cholesterol 256
    Fatty Acids 256
    N-3 PUFA Content 256
    Proximate Composition 256

Sunfish - Pumpkinseed 256
    Cholesterol 256
    Fatty Acids 256
    N-3 Content 256
    Proximate Composition 256

Swordfish 256
    Cholesterol 256
    Fatty Acids 256
    N-3 Content 256
    Proximate Composition 256

Synthesis
    Eicosanoids 29
    Leukotrienes 32
    Prostaglandins 28, 29

Thrombin
    Cod liver oil 84
    Platelet 131
    Malondialdehyde 89
    Salmon 84
    Thromboxane 85

Thrombosis 9

Thromboxane 17
    Fish oils 131

Thromboxane B3 - 9, 86
    Cod liver oil 85, 86
    Collagen 86
    Decrease 91
    Eskimos 17
    Fish oils 31, 85-87
    Lung 152
    Platelet aggregation 9, 131
    Production 131-133
    Vasoconstriction 9

Timnodonic Acid
    Fish 241-258

Tocopherol 186
    Algae 186
    Seaweed 186, 189

Trout - Artic Char 257
    Cholesterol 257
    Fatty Acids 257
    N-3 Content 257
    Proximate Composition 257

Trout - Brook 257
    Cholesterol 257
    Fatty Acids 257
    N-3 PUFA Content 257
    Proximate Composition 257

Trout - Dolly Varden 291

Trout - Lake 257, 291
    Cholesterol 257
    Fatty Acids 257
    N-3 Content 257, 291
    Habitat 291
    Landings 291
    Proximate Composition 257
    Uses 291

Trout - Rainbow 256, 292
    Cholesterol 256
    Fatty Acids 256
    N-3 PUFA Content 256, 292
    Habitat 292
    Landings 292
    Proximate Composition 256
    Uses 292

Tuna - Unspecified 257
    Cholesterol 257

Index

Fatty Acids 257
N-3 PUFA Content 257
Proximate Composition 257

Tuna - Albacore 258, 292
    Cholesterol 258
    Fatty Acids 258
    N-3 PUFA Content 258, 292
    Habitat 292
    Landings 292
    Proximate Composition 258
    Uses 292

Tuna - Bluefin 258, 293
    Cholesterol 258
    Fatty Acids 258
    N-3 PUFA Content 258, 293
    Habitat 293
    Landings 293
    Proximate Composition 258
    Uses 293

Tuna - Skipjack 258, 293
    Cholesterol 258
    Fatty Acids 258
    N-3 PUFA Content 258, 293
    Habitat 293
    Landings 293
    Proximate Composition 258
    Uses 293

Tuna - Yellowfin 258
    Cholesterol 258
    Fatty Acids 258
    N-3 Content 258
    Proximate Composition 258

Turbot 293
    Habitat 294
    Uses 294

Turbot - Arrowtooth 294
    Habitat 294
    Uses 294

Unsaturated Fatty Acids 6 see polyunsaturated
    Docosahexaenoic acid 6, 7
    Eicosapentaenoic acid 6, 7
    Linoleic acid 6,7
    Linolenic acid 6,7
    n-6,7
    n-3 6,7

Urinary
    Prostaglandins 62
    6-Keto F1alpha 91

Viscosity 94

Visual Acuity 11

Vitamins 97, 183
    A 178, 183
    D 185
    E 186

VLDL - cholesterol 77, 78
    Fatty fish 77
    Salmon oil 77

Whitefish - Lake 259, 294
    Cholesterol 259
    Fatty Acids 259
    N-3 PUFA Content 259, 294
    Habitat 294
    Landings 294
    Proximate Composition 259
    Uses 294

Wolffish - Atlantic 259
    Cholesterol 259
    Fatty Acids 259
    N-3 PUFA Content 259
    Proximate Composition 259

Whiting 259
    Cholesterol 259
    Fatty Acids 259
    N-3 PUFA Content 259
    Proximate Composition 259

Whiting - King 294